UNITED NATIONS CONFERENCE ON TRADE

# REVIEW OF MARITIME TRANSPORT

## 2018

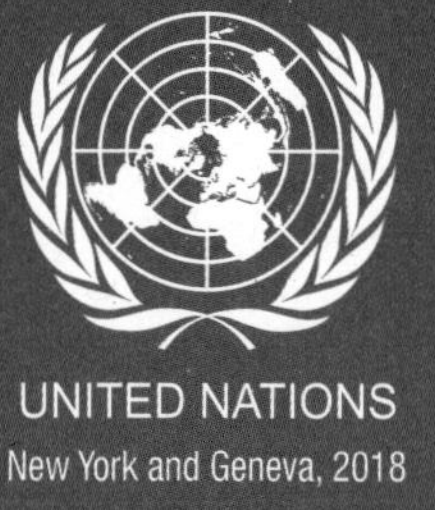

UNITED NATIONS
New York and Geneva, 2018

United Nations Publications

300 East 42nd Street

New York, New York 10017

United States of America

Email: publications@un.org

Website: un.org/publications



UNCTAD secretariat

Palais des Nations
1211 Geneva 10, Switzerland



United Nations publication issued by the United Nations Conference on Trade and Development.

UNCTAD/RMT/2018

ISBN 978-92-1-112928-1

eISBN 978-92-1-047241-8

ISSN 0566-7682

Sales No. E.18.II.D.5

# ACKNOWLEDGEMENTS

The *Review of Maritime Transport 2018* was prepared by UNCTAD under the coordination of Jan Hoffmann, with administrative support and formatting by Wendy Juan, and the overall guidance of Shamika N. Sirimanne. Regina Asariotis, Mark Assaf, Hassiba Benamara, Jan Hoffmann, Anila Premti, Luisa Rodríguez, Mathis Weller and Frida Youssef were contributing authors.

The publication was edited by the Intergovernmental Support Service of UNCTAD. The cover was designed by Magali Studer. Desktop publishing was carried out by Nathalie Loriot.

Comments and input provided by the following reviewers are gratefully acknowledged: Gail Bradford, Trevor Crowe, Neil Davidson, Mahin Faghfouri, Mike Garratt, Sarah Hutley, Katerina Konsta, Peter de Langen, Wolfgang Lehmacher, Steven Malby, Olaf Merk, James Milne, Gabriel Petrus, Harilaos N. Psaraftis, Jean-Paul Rodrigue, Tristan Smith, Antonella Teodoro and Dirk Visser.

Thanks are also due to Vladislav Shuvalov for reviewing the publication in full.

# TABLE OF CONTENTS

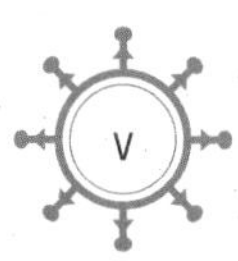


## Tables

## Figures



## Boxes

# ABBREVIATIONS

| | |
|---|---|
| **BIMCO** | Baltic and International Maritime Council |
| **dwt** | dead-weight ton(s) |
| **e-commerce** | electronic commerce |
| **FEU** | 40-foot equivalent unit |
| **GDP** | gross domestic product |
| **IBM** | International Business Machines |
| **IMO** | International Maritime Organization |
| **TEU** | 20-foot equivalent unit |

# NOTE

The *Review of Maritime Transport* is a recurrent publication prepared by the UNCTAD secretariat since 1968 with the aim of fostering the transparency of maritime markets and analysing relevant developments. Any factual or editorial corrections that may prove necessary, based on comments made by Governments, will be reflected in a corrigendum to be issued subsequently.

This edition of the Review covers data and events from January 2017 until June 2018. Where possible, every effort has been made to reflect more recent developments.

All references to dollars ($) are to United States dollars, unless otherwise stated.

"Ton" means metric ton (1,000 kg) and "mile" means nautical mile, unless otherwise stated.

Because of rounding, details and percentages presented in tables do not necessarily add up to the totals.

Two dots (..) in a statistical table indicate that data are not available or are not reported separately.

An em-dash (—) in a statistical table indicates that the amount is nil or negligible.

The terms "countries" and "economies" refer to countries, territories or areas.

Since 2014, the *Review of Maritime Transport* does not include printed statistical annexes. Instead, UNCTAD has expanded the coverage of statistical data online via the following links:

Overview: http://stats.unctad.org/maritime.

Seaborne trade: http://stats.unctad.org/seabornetrade

Merchant fleet by flag of registration: http://stats.unctad.org/fleet

Merchant fleet by country of ownership: http://stats.unctad.org/fleetownership

National maritime country profiles: http://unctadstat.unctad.org/CountryProfile/en-GB/index.html

Shipbuilding by country in which built: http://stats.unctad.org/shipbuilding

Ship scrapping by country of demolition: http://stats.unctad.org/shipscrapping

Liner shipping connectivity index: http://stats.unctad.org/lsci

Liner shipping bilateral connectivity index: http://stats.unctad.org/lsbci

Container port throughput: http://stats.unctad.org/teu

All websites cited in this report were accessed in August 2018.

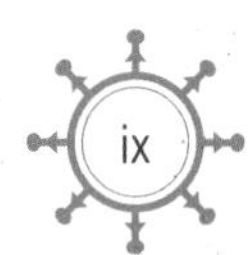


## Vessel groupings used in the *Review of Maritime Transport*

| Group | Constituent ship types |
|---|---|
| **Oil tankers** | Oil tankers |
| **Bulk carriers** | Bulk carriers, combination carriers |
| **General cargo ships** | Multi-purpose and project vessels, roll-on roll-off (ro-ro) cargo, general cargo |
| **Container ships** | Fully cellular container ships |
| **Other ships** | Liquefied petroleum gas carriers, liquefied natural gas carriers, parcel (chemical) tankers, specialized tankers, reefers, offshore supply vessels, tugs, dredgers, cruise, ferries, other non-cargo ships |
| **Total all ships** | Includes all the above-mentioned vessel types |

### Approximate vessel-size groups referred to in the *Review of Maritime Transport*, according to commonly used shipping terminology

***Crude oil tankers***

| | |
|---|---|
| Very large crude carrier | 200,000 deadweight tons (dwt) and above |
| Suezmax crude tanker | 120,000–200,000 dwt |
| Aframax crude tanker | 80,000–119,999 dwt |
| Panamax crude tanker | 60,000–79,999 dwt |

***Dry bulk and ore carriers***

| | |
|---|---|
| Capesize bulk carrier | 100,000 dwt and above |
| Panamax bulk carrier | 65,000–99,999 dwt |
| Handymax bulk carrier | 40,000–64,999 dwt |
| Handysize bulk carrier | 10,000–39,999 dwt |

***Container ships***

| | |
|---|---|
| Neo Panamax | Ships that can transit the expanded locks of the Panama Canal with up to a maximum 49 m beam and 366 m length overall |
| Panamax | Container ships above 3,000 20-foot equivalent units (TEUs) with a beam below 33.2 m, i.e. the largest size vessels that can transit the old locks of the Panama Canal |

*Source:* Clarkson Research Services.

*Note:* Unless otherwise indicated, the ships mentioned in the *Review of Maritime Transport* include all propelled seagoing merchant vessels of 100 gross tons and above, excluding inland waterway vessels, fishing vessels, military vessels, yachts, and fixed and mobile offshore platforms and barges (with the exception of floating production storage and offloading units and drillships).

# EXECUTIVE SUMMARY

## Growing seaborne trade

Global seaborne trade is doing well, supported by the 2017 upswing in the world economy. Expanding at 4 per cent, the fastest growth in five years, global maritime trade gathered momentum and raised sentiment in the shipping industry. Total volumes reached 10.7 billion tons, reflecting an additional 411 million tons, nearly half of which were made of dry bulk commodities.

Global containerized trade increased by 6.4 per cent, following the historical lows of the two previous years. Dry bulk cargo increased by 4.0 per cent, up from 1.7 per cent in 2016, while growth in crude oil shipments decelerated to 2.4 per cent. Reduced shipments from exporters of the Organization of Petroleum Exporting Countries were offset by increased trade flows originating from the Atlantic basin and moving eastward towards Asia. This new trend has reshaped crude oil trade patterns, which became less concentrated on usual suppliers from Western Asia. Supported by the growing global refining capacity – especially in Asia – and the appeal of gas as a cleaner energy source, refined petroleum products and gas increased by a combined 3.9 per cent in 2017.

Prospects for seaborne trade are positive; UNCTAD projects volume increases of 4 per cent in 2018, a rate equivalent to that of 2017. Contingent on continued favourable trends in the global economy, UNCTAD is forecasting a 3.8 per cent compound annual growth rate between 2018 and 2023. Volumes across all segments are set to grow, with containerized and dry bulk commodities expected to record the fastest growth at the expense of tanker volumes. UNCTAD projections for overall seaborne trade are consistent with historical trends, whereby seaborne trade increased at an annual average rate of 3.5 per cent between 2005 and 2017. Projections of rapid growth in dry cargo are in line with a five-decade-long pattern that saw the share of tanker volumes being displaced by dry cargoes, dropping from over 50 per cent in 1970 to less than 33 per cent in 2017.

## Uncertain outlook

While the prospects for seaborne trade are bright, downside risks such as increased inward-looking policies and the rise of trade protectionism are, nevertheless, weighing on the outlook. An immediate concern is the trade tensions between China and the United States of America, the world's two largest economies, as well as those between Canada, Mexico, the United States and the European Union. Escalating trade frictions may lead to a trade war that could derail recovery, reshape global maritime trade patterns and dampen the outlook. Further, there are other factors driving uncertainty. Among others, these include the ongoing global energy transition, structural shifts in economies such as China, and shifts in global value chain development patterns.

If leveraged effectively, game-changing trends, such as digitalization, electronic commerce (e-commerce) and the Belt and Road Initiative, the exact impact of which is yet to be fully understood, have the potential to add wind to the sails of global seaborne trade.

## Growth in world fleet capacity

After five years of decelerating growth, 2017 saw a small improvement in world fleet expansion. During the year, a total of 42 million gross tons were added to global tonnage, equivalent to a 3.3 per cent growth rate. This performance reflects both a slight upturn in new deliveries and a decline in demolition activity, except in the tanker market, where demolition activity picked up. The expansion in ship supply capacity was surpassed by faster growth in seaborne trade volumes, altering the market balance and supporting improved freight rates and earnings.

With regard to the shipping value chain, Germany remained the largest container shipowning country, although it lost some ground in 2017. In contrast, owners from Canada, China and Greece expanded their containership-owning market shares. The Marshall Islands emerged as the second-largest registry, after Panama and ahead of Liberia. Over 90 per cent of shipbuilding activity occurred in China, Japan and the Republic of Korea, while 79 per cent of ship demolitions took place in South Asia, notably in Bangladesh, India and Pakistan.

## Improved balance between demand and supply

Supported by stronger global demand, more manageable fleet-capacity growth and overall better market conditions, freight rate levels improved significantly in 2017, except for those of the tanker market. Container freight rate levels increased, with averages surpassing performance in 2016 and with profits in the container shipping industry reaching roughly $7 billion by the end of 2017. CMA CGM recorded the best operating results in the container shipping industry, with core earnings before interest and taxes reaching close to $1.58 billion, followed by Maersk Line, with gains of $700 million. Hapag-Lloyd ranked third, with gains amounting to some $480 million. The 2017 surge in bulk freight market resulted in gains for carriers that helped offset the depressed earnings of 2016. The tanker market remained under pressure, owing mainly to increased vessel supply capacity that outpaced demand growth and undermined freight rates.

While these trends are positive for shipping, recovery remains nevertheless fragile in view of the highly volatile rates yet relatively low levels.

## Consolidation activity in liner shipping

The liner shipping industry witnessed further consolidation through mergers and acquisitions and global alliance restructuring. Yet despite the global market concentration trend, UNCTAD observed growth in the average number of companies providing services per country between 2017 and 2018. This is the first increase since UNCTAD started monitoring capacity deployment in 2004. Put differently, several individual carriers – both inside and outside alliances – expanded their networks to a larger number of countries. This more than offset the reduction in the global number of companies after the takeovers and mergers. However, this was not a broad-based trend. The number of operators servicing several small island developing States and vulnerable economies decreased between 2017 and 2018.

Three global liner shipping alliances dominate capacity deployed on the three major East–West container routes, collectively accounting for 93 per cent of deployed capacity. Alliance members continue to compete on price while operational efficiency and capacity utilization gains are helping to maintain low freight rate levels. By joining forces and forming alliances, carriers have strengthened their bargaining power vis-à-vis seaports when negotiating port calls and terminal operations.

In an oversupplied market, consolidation is expected to continue. Two thirds of the container ship order book capacity is accounted for by ships of over 14,000 TEUs, and only large carriers and alliances are in a position to fill these mega ships.

## Port traffic volumes

Global port activity and cargo handling expanded rapidly in 2017, following two years of weak performance. According to 2017 estimates, the top 20 global ports handled 9.3 billion tons, up from 8.9 billion tons in 2016, an amount nearly equivalent to global seaborne trade volumes. UNCTAD estimates that 752.2 million TEUs were moved at container ports worldwide in 2017. This total reflects the addition of some 42.3 million TEUs in 2017, an amount comparable to total container volumes handled that year by the world busiest container port, Shanghai, China.

The outlook for global port-handling activity remains positive overall, supported by projected economic growth and port infrastructure development plans. However, downside risks weighing on global demand and related uncertainty continue to diminish global port activity.

## Port operations, performance and bargaining power

Liner shipping alliances and vessel upsizing have made the relationship between container shipping lines and ports more complex and have triggered new dynamics where shipping lines have greater bargaining power and influence. Vessel size increases and the rise of mega alliances have heightened the requirements for ports to adapt. While liner shipping networks seem to have benefited from efficiency gains arising from consolidation and alliance restructuring, the benefits for ports have not evolved at the same pace.

Together, these trends have heightened competition among container ports to win port calls with decisions by shipping alliances regarding capacity deployed, ports of call and network structures being potentially able to determine the fate of a container port terminal. This dynamic is further complicated by the shipping lines often being involved in port operations, which in turn could redefine approaches to terminal concessions.

## Tracking and measuring port performance for strategic planning and decision-making

Global ports and terminals need to track and measure performance, as port performance metrics enable sound strategic planning and decision-making, as well as informed investment and financing decisions. As global trade, supply chains, production processes and countries' effective integration into the world economy are heavily dependent on well-functioning port systems, it is becoming increasingly important to monitor and measure the operational, financial, economic, environmental and social performance of ports.

In this respect, improved data availability enabled by various technological advances can be tapped. In addition, work carried out under the UNCTAD Port Management Programme and the port performance scorecard could be further strengthened.

## Challenges and opportunities of digitalization

Technological advances in the shipping industry, such as autonomous ships, drones and various blockchain applications, hold considerable promise for the supply side of shipping. However, there is still uncertainty within the maritime industry regarding possible safety, security and cybersecurity incidents, as well as concern about negative effects on the jobs of seafarers, most of which come from developing countries.

While the development and use of autonomous ships offer numerous benefits, it is still unclear whether this new technology will be fully accepted by Governments, and particularly by the traditionally conservative maritime industry. There are legitimate concerns about the safety and security of operation of autonomous ships and their reliability. The diminishing role of seafarers and ensuing job loss are a particular concern.

At present, many blockchain technology initiatives and partnerships have the potential to be used for tracking cargo and providing end-to-end supply chain visibility;

recording information on vessels, including on global risks and exposures; integrating smart contracts and marine insurance policies; and digitalizing and automating paper filings and documents, thus saving time and cost for clearance and movement of cargo. Combining on-board systems and digital platforms allow for vessels and their cargo to become part of the Internet of things. A key challenge will be to establish interoperability so that data can be exchanged seamlessly, while ensuring at the same time cybersecurity and the protection of commercially sensitive or private data, including in view of the recent General Data Protection Regulation of the European Union. [1]

Many technological advances are applicable in ports and terminals and offer an opportunity for port stakeholders to innovate and generate additional value in the form of greater efficiency, enhanced productivity, greater safety and heightened environmental protection. In light of these developments, ports and terminals worldwide need to re-evaluate their role in global maritime logistics and prepare to effectively embrace and leverage digitalization-driven innovations and technologies.

## International shipping commitment to reduced greenhouse gas emissions

Complementing international efforts to address greenhouse gas emissions, which include the Paris Agreement under the United Nations Framework Convention on Climate Change and the 2030 Agenda for Sustainable Development, in particular Sustainable Development Goal 13 to take urgent action to combat climate change and its impacts, an important achievement was made at the International Maritime Organization (IMO) related to the determination of international shipping's fair share of greenhouse gas emissions reduction. An initial strategy on the reduction of such emissions from ships was adopted in April 2018, according to which total annual greenhouse gas emissions would be reduced by at least 50 per cent by 2050, compared with 2008. The strategy identifies short-, medium- and long-term further measures with possible timelines, and their impacts on States, paying particular attention to the needs of developing countries, especially small island developing States and the least developed countries. It also identifies supportive measures, including capacity-building, technical cooperation, and research and development. Innovative emissions reduction mechanisms, possibly including market-based measures, are proposed as medium-term solutions to be decided upon between 2023 and 2030, along with possible long-term measures to be undertaken beyond 2030.

Related regulatory developments of note include the entry into force of amendments to the International Convention for the Prevention of Pollution from Ships, 1973/1978, to make mandatory the data collection system for fuel oil consumption of ships of 5,000 gross tons and above; data collection is required to start as of 1 January 2019. As regards ship-source air pollution, associated with a large number of respiratory illnesses and deaths, the global limit of 0.5 per cent on sulphur in fuel oil used on board ships will come into effect on 1 January 2020, with potentially important benefits for human health and the environment. To facilitate and support effective implementation of the global limit, relevant guidelines are under preparation at IMO.

## Key trends shaping the outlook

The Review has identified seven key trends that are currently redefining the maritime transport landscape and shaping the sector's outlook. They entail the following challenges and opportunities, which require continued monitoring and assessment for sound and effective policymaking:

- First, on the demand side, the uncertainty arising from wide-ranging geopolitical, economic, and trade policy risks, as well as some structural shifts, have a negative impact on maritime trade. Of immediate concern are inward-looking policies and rising protectionist sentiment that could undermine global economic growth, restrict trade flows and shift their patterns.
- Second, the continued unfolding of digitalization and e-commerce and the implementation of the Belt and Road Initiative. These bear major implications for shipping and maritime trade.
- Third, from the supply-side perspective, overly optimistic carriers competing for market share may order excessive new capacity, thereby leading to worsened shipping market conditions. This, in turn, will upset the supply and demand balance and have repercussions on freight-rate levels and volatility, transport costs and earnings.
- Fourth, liner shipping consolidation through mergers and alliances has been on the rise in recent years in response to lower demand levels and oversupplied shipping capacity dominated by mega container ships. The implication for competition levels, the potential for market power abuse by large shipping lines and the related impact on smaller players remain a concern. Competition authorities and regulators, as well as other relevant entities such as UNCTAD, need to remain vigilant. In this respect, the seventeenth session of the Intergovernmental Group of Experts on Competition Law and Policy of UNCTAD, held in

1. Regulation (EU) 2016/679 of the European Parliament and of the Council of 27 April 2016 on the protection of natural persons with regard to the processing of personal data and on the free movement of such data, and repealing Directive 95/46/EC.

Geneva, Switzerland, in July 2018, included a round-table discussion on challenges in competition and regulation faced by developing countries in the maritime transport sector. This provided a timely opportunity to bring together competition authority representatives and other stakeholders from the sector to reflect upon some of these concerns and assess their extent and the potential implications for competition, shipping, ports and seaborne trade, as well as the role of competition law and policy in addressing these concerns. The Intergovernmental Group of Experts called upon UNCTAD to continue its analytical work in the area of international maritime transport, including the monitoring and analysis of the effects of cooperative arrangements and mergers, not only on freight rates but also on the frequency, efficiency, reliability and quality of services.

- Fifth, alliance restructuring and larger vessel deployment are also redefining the relationship between ports and container shipping lines. Competition authorities and maritime transport regulators should also analyse the impact of market concentration and alliance deployment on the relationship between ports and carriers. Areas of interest include the selection of ports of call, the configuration of liner shipping networks, the distribution of costs and benefits between container shipping and ports, and approaches to container terminal concessions.
- Sixth, the value of shipping can no longer be determined by scale alone. The ability of the sector to leverage relevant technological advances is becoming increasingly important.
- Finally, efforts to curb the carbon footprint and improve the environmental performance of international shipping remain high on the international agenda. In April 2018, IMO adopted an initial strategy to reduce annual greenhouse gas emissions from ships by at least 50 per cent by 2050 compared with 2008 – a particularly important development. With regard to air pollution, the global limit of 0.5 per cent on sulphur in fuel oil used on board ships will come into effect on 1 January 2020. To ensure consistent implementation of the global cap on sulphur, it will be important for shipowners and operators to continue to consider and adopt various strategies, including installing scrubbers and switching to liquefied natural gas and other low sulphur fuels.

# 1

# DEVELOPMENTS IN INTERNATIONAL SEABORNE TRADE

World seaborne trade gathered momentum in 2017, with volumes expanding at 4 per cent, the fastest growth in five years. Supported by the world economic recovery and the improved global merchandise trade, world seaborne trade was estimated at 10.7 billion tons, with dry bulk commodities powering nearly half of the volume increase. Bearing in mind the low base effect, the recovery benefited all market segments; containerized trade and dry bulk commodities recorded the fastest expansion. Following the weak performances of the two previous years, containerized trade increased by 6.4 per cent in 2017. Meanwhile, dry bulk commodities trade increased by 4.0 per cent, up from 1.7 per cent in 2016. Crude oil shipments rose by 2.4 per cent, down from 4 per cent in 2016, while, together, refined petroleum products and gas increased by an estimated 3.9 per cent.

UNCTAD analysis is pointing to continued growth in world seaborne trade that hinges on the continued improvement of the global economy. In line with projected growth in world gross domestic product (GDP), UNCTAD expects global maritime trade to grow by another 4 per cent in 2018. Further, world seaborne trade is projected to expand at a compound annual growth rate of 3.8 per cent between 2018 and 2023. Volumes across all segments are set to grow, with containerized and dry bulk commodities trades recording the best performances. Tanker trade volumes are also projected to increase, although at a slightly slower pace than other market segments, a trend that is consistent with historical patterns.

Although prospects for seaborne trade are positive, caution would be advisable, given the uncertainty surrounding the sustainability of the recovery and related implications for shipping. Much of the uncertainty derives from the confluence of geopolitical, economic and trade policy risks and structural shifts such as the rebalancing of the Chinese economy, slower growth of global value chains and changes in the global energy mix. This is further amplified by the emergence of new trends, notably digitalization, which could alter the face of global shipping and redefine seaborne trade flows and patterns. How these factors will evolve and the extent to which they will support or derail the recovery in seaborne trade, remains unclear. What is clear is that they will require further monitoring and assessment.

## WORLD SEABORNE TRADE IN 2017

BY VOLUME!

Global volumes gathered momentum and reached **10.7 billion tons.**

**4% annual growth:** fastest growth in five years

Containerized trade accounted for 17.1% of total seaborne trade +6.4%.

Major dry bulk commodities accounted for 29.9% of total seaborne trade +5.1%.

Crude oil shipments rose by 2.4% down from 4% in 2016.

Combined volumes of refined petroleum products and gas went up by 3.9%.

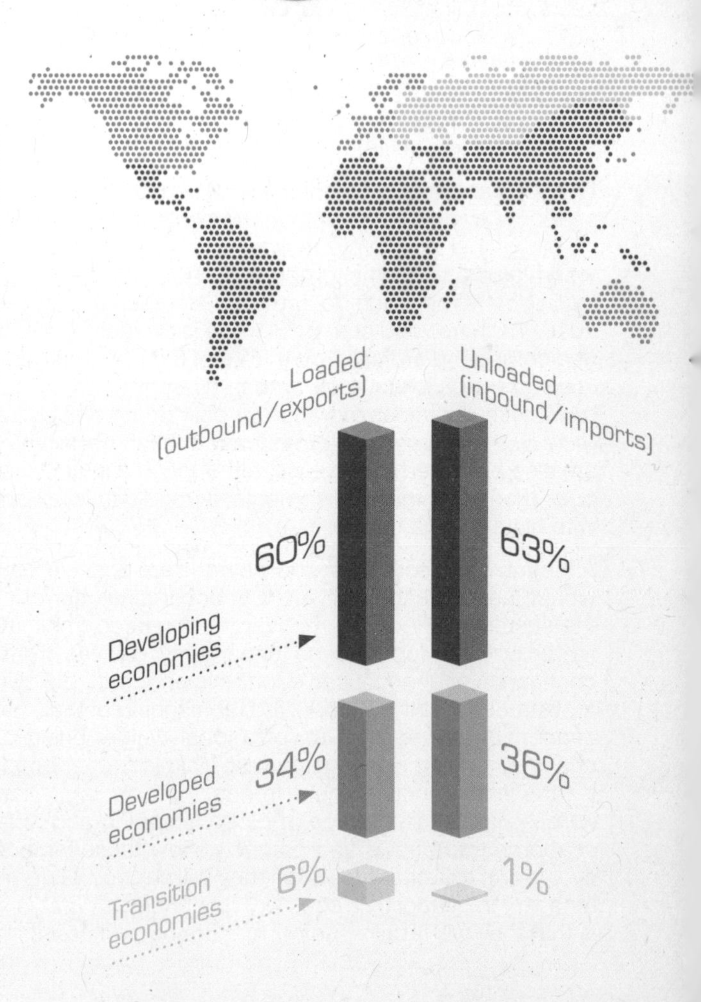


## WORLD SEABORNE TRADE GROWTH FORECAST: 2018–2023

Volume projected to grow +3.8%

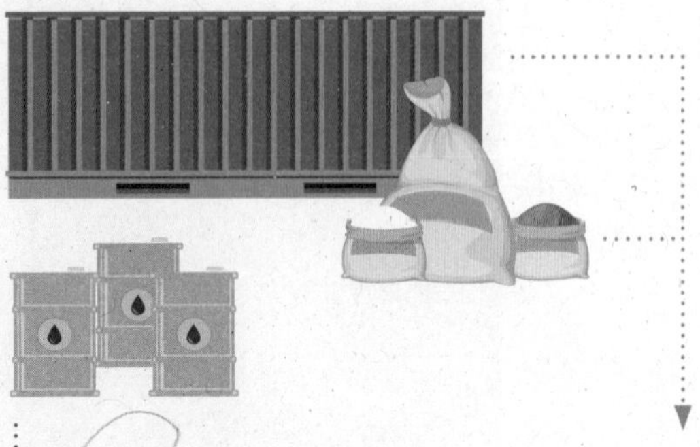

Volumes across all segments set to grow: **containerized** and **dry bulk** cargoes projected to grow the **fastest**

**Tanker** volumes to grow at a slower pace

## A. GENERAL TRENDS

Global economic expansion is the main driver of world shipping demand, and 2017 will be remembered as the year when the world economy and global shipping experienced a cyclical recovery from the historic lows of 2016, nearly a decade after the 2008–2009 global economic and financial crisis. Main economic and shipping indicators trended upward, reflecting growth in global investment, manufacturing activity and merchandise trade. At the same time, a range of upside and downside risks continued to unfold, bringing major implications for shipping and maritime trade.

### 1. Improved market fundamentals

Global industrial activity and manufacturing improved in 2017. In countries of the Organization for Economic Cooperation and Development, industrial production increased by 2.8 per cent, up from 0.2 per cent in 2016. Industrial activity in developing regions also picked up. In China, industrial production, at 6.5 per cent, was up, compared with 6 per cent in 2016. In Brazil, industrial production recovered and rose by 2.4 per cent, following the 6.4 per cent contraction recorded during the 2016 recession.

With GDP expanding by 3.1 per cent in 2017, up from 2.5 per cent in 2016, the global economy experienced a broad upswing, generating positive impacts on seaborne trade (table 1.1). Driven largely by stronger capital spending and global demand, GDP in developed countries increased by 2.3 per cent, up from 1.7 per cent in 2016. While growth accelerated in all major economies, strong growth in the European Union (2.4 per cent) was a welcome development. Growth in developing countries accelerated to 4.5 per cent, compared with 3.9 per cent in 2016, reflecting, among other factors, improved activity in commodity-exporting countries and a more favourable economic environment. This was illustrated by a return to positive growth in developing America, coinciding with the end of the recession in Brazil. A similar trend was observed in transition economies. These economies experienced positive growth in 2017, following the end of the recession in the Russian Federation. Aggregated GDP growth of 4.3 per cent in the least developed countries has improved, although it is still below the 7 per cent annual GDP growth target of the Sustainable Development Goals.

In addition to GDP, heightened global trade activity further supported maritime trade. In 2017, international merchandise trade volumes expanded by 4.7 per cent, up from 1.8 per cent in 2016 (table 1.2). Merchandise trade volumes increased in line with positive trends in the world economy, an upturn in investment and the rise in commodity prices. Higher commodity price levels translated into improved export earnings of commodity-exporting countries, which in turn, helped support their demand for imports. Rapid trade growth reflected to a large extent the trade correlation between investment and capital spending on the one hand, and merchandise trade on the other. Generally, investment tends to be more import intensive compared with other components of aggregate demand. On average, the import content of investment is estimated at about 30 per cent globally, while for private consumption and government spending, import content hovers around 23 per cent and 15 per cent, respectively (International Monetary Fund, 2016). Accelerated investment growth has thus been particularly beneficial for shipping and maritime trade, in particular for dry bulk commodities and containerized trade.

Rapid trade growth increased trade-income elasticity. The ratio of trade growth to GDP growth increased from 0.7 in 2016 to 1.7 in 2017. Nevertheless, this ratio remains low compared with the elasticities observed in the 1990s and early 2000s. As stated in previous editions of the *Review of Maritime Transport*, structural factors weighing down on trade growth also seem to be at play, along with cyclical drivers (UNCTAD, 2016).

Certain regional variations between imports and exports, as well as between country groupings, shaped trade patterns in 2017. While export growth accelerated in both the developed and developing regions, trade volumes of developing countries firmed up. Their import demand increased by 7.2 per cent, up from 1.9 per cent in 2016. Their exports expanded at 5.7 per cent, higher than the 2.3 per cent recorded in 2016. Exports from developing Asia in particular strengthened during the year following a rebound in electrical and electronic products trade and the region's integration in global value chains.

Asia recorded the fastest growth in exports (6.7 per cent) and imports (9.6 per cent). Stronger domestic Asian demand supported by policy stimulus measures in countries such as China have sustained the region's demand for imports. Developments in China are of acute relevance to shipping, as the country remained at the centre of shipping activity in 2017 and accounted for nearly half of seaborne trade growth recorded during the year.

An important development in China, which had implications for shipping and maritime trade – in particular, dry bulk shipping – was the rapid expansion of the country's GDP (6.9 per cent), reflecting a short-term deviation from the gradual rebalancing of its economy towards services and domestic consumption. Another shift observed in 2017 was the growing focus on controlling air pollution in China and related implications for the energy mix, the quality of raw materials sourced and the domestic production versus import trade-offs. These trends favoured the sourcing of commodities of better quality or grades from external markets, which in turn, contributed to boosting import volumes in China.

Demand for imports improved markedly in developing America, following negative growth in 2016. Large economies such as Argentina and Brazil, which emerged from the recession in 2017, achieved positive results. In contrast, demand for imports in Africa,

Western Asia and transition economies remained under pressure (0.9 per cent growth in 2017), despite some improvement over 2016. Among other factors, this was a reflection of the continued weakness of commodity prices and exports, and the impact of the recession in the Russian Federation.

Demand for imports in the developed regions strengthened; volumes expanded by 3.1 per cent in 2017, compared with 2 per cent in 2016. Merchandise export volumes in these regions increased by 3.5 per cent, up from 1.1 per cent in 2016.

**Table 1.1 World economic growth, 2016–2018**
(Annual percentage change)

| Region or country | 2016 | 2017[a] | 2018[b] |
|---|---|---|---|
| World | 2.5 | 3.1 | 3.0 |
| Developed countries | 1.7 | 2.3 | 2.1 |
| of which: | | | |
| United States | 1.5 | 2.3 | 2.5 |
| European Union (28) | 2.0 | 2.6 | 2.0 |
| Japan | 1.0 | 1.7 | 0.9 |
| Developing countries | 3.9 | 4.5 | 4.6 |
| of which: | | | |
| Africa | 1.7 | 3.0 | 3.5 |
| East Asia | 5.9 | 6.2 | 6.0 |
| of which: | | | |
| China | 6.7 | 6.9 | 6.7 |
| South Asia | 8.4 | 5.8 | 6.1 |
| of which: | | | |
| India | 7.9 | 6.2 | 7.0 |
| Western Asia | 3.1 | 3.0 | 3.3 |
| Latin American and the Caribbean | -1.1 | 1.1 | 1.8 |
| of which: | | | |
| Brazil | -3.5 | 1.0 | 1.4 |
| Countries with economies in transition | 0.3 | 2.1 | 2.2 |
| of which: | | | |
| Russian Federation | -0.2 | 1.5 | 1.7 |
| Least developed countries | 3.5 | 4.3 | 4.9 |

*Source:* UNCTAD secretariat calculations, based on United Nations, 2018 and UNCTAD, 2018a.
[a] Partly estimated.
[b] Forecast.

## 2. Growing world seaborne trade

International seaborne trade gathered momentum, with volumes expanding by 4 per cent. This was the fastest growth in five years. Reflecting the world economic recovery and improved global merchandise trade, UNCTAD estimates world seaborne trade volumes at 10.7 billion tons in 2017 (tables 1.3 and 1.4, figure 1.1). Dry bulk commodities have powered nearly half of the volume increase.

Major dry bulk commodities – coal, iron ore and grain – accounted for 42.3 per cent of total dry cargo shipments, which were estimated at 7.6 billion tons in 2017. Containerized trade and minor bulks represented 24.3 per cent and 25.4 per cent of the total, respectively. Remaining volumes were made of other dry cargo, including breakbulk shipments.

Tanker trade shipments accounted for less than one third of total seaborne trade volume, in line with the persistent shift in the structure of seaborne trade observed over the past four decades. The share of tanker trade dropped from around 55 per cent in 1970 to 29.4 per cent in 2017. Between 1980 and 2017, global tanker trade expanded at an annual average growth rate of 1.4 per cent, while major dry bulks rose by 4.6 per cent. The fastest growing segment was containerized trade, with volumes expanding over nearly four decades at an annual average growth rate of 8.1 per cent.

Developing countries continue to account for most global seaborne trade flows, both in terms of exports (goods loaded) and imports (goods unloaded). These countries shipped 60 per cent of world merchandise trade by sea in 2017 and unloaded 63 per cent of this total. By contrast, developed countries saw their share of both types of traffic decline over the years, representing about one third of world seaborne imports and exports (34 per cent of goods loaded and 36 per cent, unloaded). Transition economies continue to be heavily reliant on the export of bulky raw materials and commodities (6 per cent), while they hold a marginal share of global seaborne imports (1 per cent).

**Table 1.2 Growth in volume of merchandise trade, 2015–2017**
(Annual percentage change)

| Exports | | | Countries or regions | Imports | | |
|---|---|---|---|---|---|---|
| 2015 | 2016 | 2017 | | 2015 | 2016 | 2017 |
| 2.5 | 1.8 | 4.7 | World[a] | 2.5 | 1.8 | 4.7 |
| 2.3 | 1.1 | 3.5 | Developed countries | 4.3 | 2.0 | 3.1 |
| 2.4 | 2.3 | 5.7 | Developing countries | 0.6 | 1.9 | 7.2 |
| 0.8 | 0.6 | 4.2 | North America | 5.4 | 0.1 | 4.0 |
| 1.8 | 1.9 | 2.9 | Latin America and the Caribbean | -6.4 | -6.8 | 4.0 |
| 2.9 | 1.1 | 3.5 | Europe | 3.7 | 3.1 | 2.5 |
| 1.5 | 2.3 | 6.7 | Asia | 4.0 | 3.5 | 9.6 |
| 5.5 | 2.6 | 2.3 | Africa, Western Asia and countries with economies in transition | -5.6 | 0.2 | 0.9 |

*Source:* UNCTAD secretariat, based on World Trade Organization, 2018, table 1.
[a] Average of exports and imports.

**Table 1.3 Development in international seaborne trade, selected years**
(Millions of tons loaded)

| Year | Crude oil, petroleum products and gas | Main bulks[a] | Other dry cargo[a] | Total (all cargoes) |
|---|---|---|---|---|
| 1970 | 1 440 | 448 | 717 | 2 605 |
| 1980 | 1 871 | 608 | 1 225 | 3 704 |
| 1990 | 1 755 | 988 | 1 265 | 4 008 |
| 2000 | 2 163 | 1 295 | 2 526 | 5 984 |
| 2005 | 2 422 | 1 711 | 2 976 | 7 109 |
| 2006 | 2 698 | 1 713 | 3 289 | 7 701 |
| 2007 | 2 747 | 1 840 | 3 447 | 8 034 |
| 2008 | 2 742 | 1 946 | 3 541 | 8 229 |
| 2009 | 2 642 | 2 022 | 3 194 | 7 858 |
| 2010 | 2 772 | 2 259 | 3 378 | 8 409 |
| 2011 | 2 794 | 2 392 | 3 599 | 8 785 |
| 2012 | 2 841 | 2 594 | 3 762 | 9 197 |
| 2013 | 2 829 | 2 761 | 3 924 | 9 514 |
| 2014 | 2 825 | 2 988 | 4 030 | 9 843 |
| 2015 | 2 932 | 2 961 | 4 131 | 10 024 |
| 2016 | 3 055 | 3 041 | 4 193 | 10 289 |
| 2017 | 3 146 | 3 196 | 4 360 | 10 702 |

*Source:* UNCTAD secretariat calculations, based on data supplied by reporting countries and as published on government and port industry websites, and by specialist sources.

*Notes:* Dry cargo data for 2006 onwards were revised and updated to reflect improved reporting, including more recent figures and a better breakdown by cargo type. Since 2006, the breakdown of dry cargo into main bulks and dry cargo other than main bulks is based on various issues of the *Shipping Review and Outlook*, produced by Clarksons Research. Total estimates of seaborne trade figures for 2017 are based on preliminary data or on the last year for which data were available.

[a] *Figures* for main bulks include data on iron ore, grain, coal, bauxite/alumina and phosphate. Starting in 2006, they include data on iron ore, grain and coal only. Data relating to bauxite/alumina and phosphate are included under "other dry cargo".

**Figure 1.1 International seaborne trade, selected years**
(Millions of tons loaded)

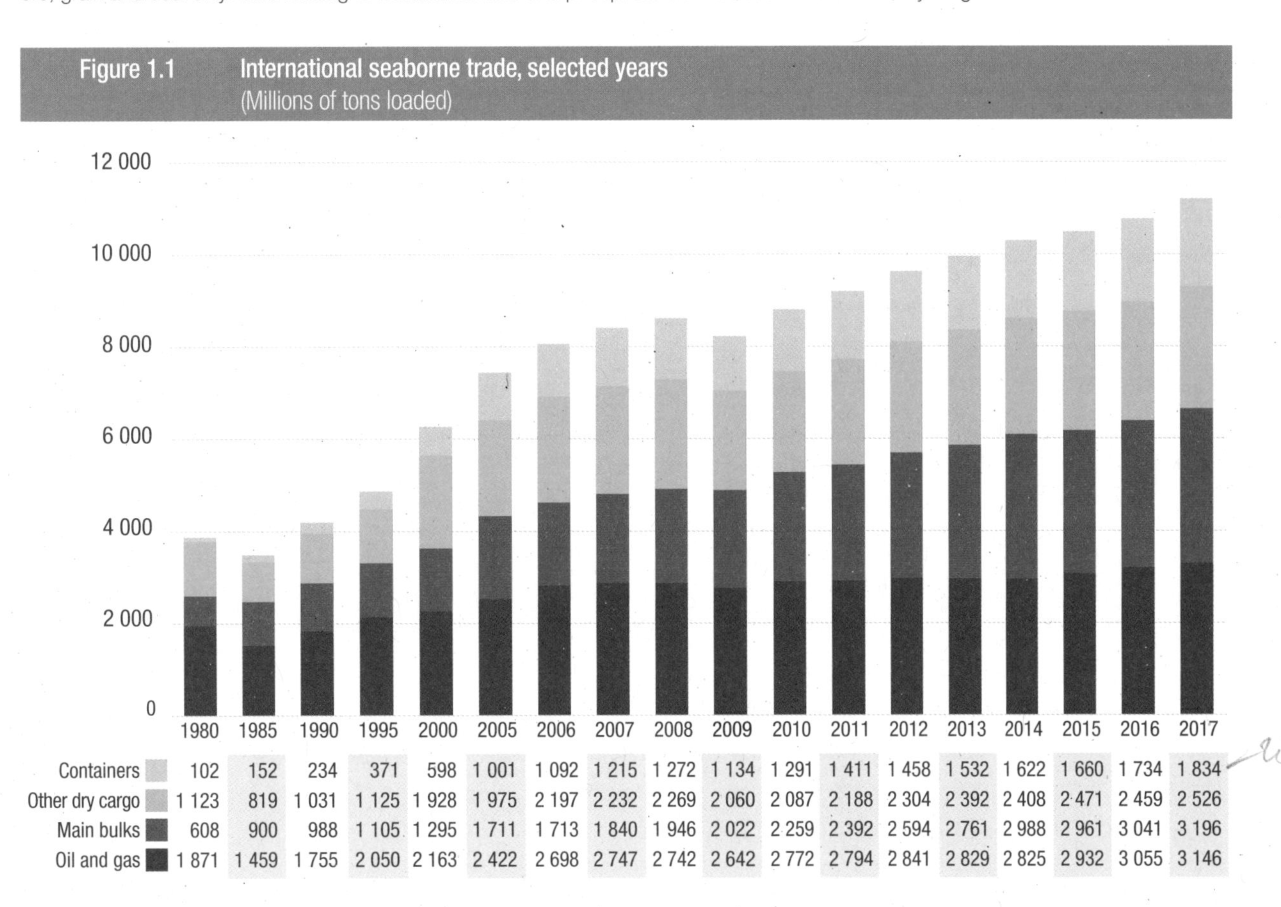


| | 1980 | 1985 | 1990 | 1995 | 2000 | 2005 | 2006 | 2007 | 2008 | 2009 | 2010 | 2011 | 2012 | 2013 | 2014 | 2015 | 2016 | 2017 |
|---|---|---|---|---|---|---|---|---|---|---|---|---|---|---|---|---|---|---|
| Containers | 102 | 152 | 234 | 371 | 598 | 1 001 | 1 092 | 1 215 | 1 272 | 1 134 | 1 291 | 1 411 | 1 458 | 1 532 | 1 622 | 1 660 | 1 734 | 1 834 |
| Other dry cargo | 1 123 | 819 | 1 031 | 1 125 | 1 928 | 1 975 | 2 197 | 2 232 | 2 269 | 2 060 | 2 087 | 2 188 | 2 304 | 2 392 | 2 408 | 2 471 | 2 459 | 2 526 |
| Main bulks | 608 | 900 | 988 | 1 105 | 1 295 | 1 711 | 1 713 | 1 840 | 1 946 | 2 022 | 2 259 | 2 392 | 2 594 | 2 761 | 2 988 | 2 961 | 3 041 | 3 196 |
| Oil and gas | 1 871 | 1 459 | 1 755 | 2 050 | 2 163 | 2 422 | 2 698 | 2 747 | 2 742 | 2 642 | 2 772 | 2 794 | 2 841 | 2 829 | 2 825 | 2 932 | 3 055 | 3 146 |

*Source: Review of Maritime Transport*, various issues. For 2006–2017, the breakdown by cargo type is based on Clarksons Research, 2018a.

*Notes:* 1980–2005 figures for main bulks include iron ore, grain, coal, bauxite/alumina and phosphate. Starting in 2006, main bulks include iron ore, grain and coal only. Data relating to bauxite/alumina and phosphate are included under "other dry cargo".

**Table 1.4 World seaborne trade, 2016–2017**
(Type of cargo, country group and region)

| Country group | Year | Goods loaded | | | | Goods unloaded | | | |
|---|---|---|---|---|---|---|---|---|---|
| | | Total | Crude oil | Petroleum products and gas | Dry cargo | Total | Crude oil | Petroleum products and gas | Dry cargo |
| | | Millions of tons | | | | | | | |
| World | 2016 | 10 288.6 | 1 831.4 | 1 223.7 | 7 233.5 | 10 279.9 | 1 990.0 | 1 235.7 | 7 054.1 |
| | 2017 | 10 702.1 | 1 874.9 | 1 271.2 | 7 555.9 | 10 666.0 | 2 035.0 | 1 281.5 | 7 349.4 |
| Developed economies | 2016 | 3 492.9 | 150.5 | 453.0 | 2 889.4 | 3 840.4 | 1 001.3 | 507.6 | 2 331.5 |
| | 2017 | 3 675.0 | 162.6 | 478.3 | 3 034.2 | 3 838.3 | 956.8 | 509.1 | 2 372.5 |
| Transition economies | 2016 | 637.3 | 176.3 | 40.2 | 420.7 | 59.6 | 0.3 | 4.0 | 55.3 |
| | 2017 | 664.5 | 190.7 | 48.3 | 425.6 | 65.9 | 0.8 | 3.4 | 61.7 |
| Developing economies | 2016 | 6 158.4 | 1 504.5 | 730.5 | 3 923.4 | 6 379.9 | 988.5 | 724.2 | 4 667.3 |
| | 2017 | 6 362.5 | 1 521.6 | 744.7 | 4 096.2 | 6 761.7 | 1 077.4 | 769.1 | 4 915.3 |
| Africa | 2016 | 692.7 | 271.3 | 58.8 | 362.6 | 492.9 | 38.7 | 80.8 | 373.4 |
| | 2017 | 726.2 | 288.0 | 60.0 | 378.2 | 499.8 | 33.9 | 90.5 | 375.4 |
| America | 2016 | 1 336.8 | 232.5 | 75.9 | 1 028.4 | 566.0 | 51.9 | 128.2 | 385.8 |
| | 2017 | 1 379.4 | 227.3 | 71.9 | 1 080.2 | 608.3 | 54.7 | 141.8 | 411.8 |
| Asia | 2016 | 4 121.2 | 999.1 | 594.9 | 2 527.2 | 5 307.6 | 897.0 | 510.9 | 3 899.7 |
| | 2017 | 4 248.8 | 1 004.6 | 611.8 | 2 632.4 | 5 640.1 | 988.0 | 532.5 | 4 119.6 |
| Oceania | 2016 | 7.7 | 1.7 | 0.9 | 5.2 | 13.5 | 0.8 | 4.2 | 8.4 |
| | 2017 | 8.0 | 1.7 | 0.9 | 5.4 | 13.5 | 0.8 | 4.2 | 8.4 |
| **Country group** | **Year** | **Goods loaded** | | | | **Goods unloaded** | | | |
| | | **Total** | **Crude oil** | **Petroleum products and gas** | **Dry cargo** | **Total** | **Crude oil** | **Petroleum products and gas** | **Dry cargo** |
| | | Percentage share | | | | | | | |
| World | 2016 | 100.0 | 17.8 | 11.9 | 70.3 | 100.0 | 19.4 | 12.0 | 68.6 |
| | 2017 | 100.0 | 17.5 | 11.9 | 70.6 | 100.0 | 19.1 | 12.0 | 68.9 |
| Developed economies | 2016 | 33.9 | 8.2 | 37.0 | 39.9 | 37.4 | 50.3 | 41.1 | 33.1 |
| | 2017 | 34.3 | 8.7 | 37.6 | 40.2 | 36.0 | 47.0 | 39.7 | 32.3 |
| Transition economies | 2016 | 6.2 | 9.6 | 3.3 | 5.8 | 0.6 | 0.0 | 0.3 | 0.8 |
| | 2017 | 6.2 | 10.2 | 3.8 | 5.6 | 0.6 | 0.0 | 0.3 | 0.8 |
| Developing economies | 2016 | 59.9 | 82.2 | 59.7 | 54.2 | 62.1 | 49.7 | 58.6 | 66.2 |
| | 2017 | 59.5 | 81.2 | 58.6 | 54.2 | 63.4 | 52.9 | 60.0 | 66.9 |
| Africa | 2016 | 6.7 | 14.8 | 4.8 | 5.0 | 4.8 | 1.9 | 6.5 | 5.3 |
| | 2017 | 6.8 | 15.4 | 4.7 | 5.0 | 4.7 | 1.7 | 7.1 | 5.1 |
| America | 2016 | 13.0 | 12.7 | 6.2 | 14.2 | 5.5 | 2.6 | 10.4 | 5.5 |
| | 2017 | 12.9 | 12.1 | 5.7 | 14.3 | 5.7 | 2.7 | 11.1 | 5.6 |
| Asia | 2016 | 40.1 | 54.6 | 48.6 | 34.9 | 51.6 | 45.1 | 41.3 | 55.3 |
| | 2017 | 39.7 | 53.6 | 48.1 | 34.8 | 52.9 | 48.5 | 41.6 | 56.1 |
| Oceania | 2016 | 0.1 | 0.1 | 0.1 | 0.1 | 0.1 | 0.0 | 0.3 | 0.1 |
| | 2017 | 0.1 | 0.1 | 0.1 | 0.1 | 0.1 | 0.0 | 0.3 | 0.1 |

*Source:* UNCTAD secretariat calculations, based on data supplied by reporting countries and as published on government and port industry websites, and by specialist sources.

*Notes:* Dry cargo data for 2006 onwards were revised and updated to reflect improved reporting, including more recent figures and a better breakdown by cargo type. Total estimates of seaborne trade figures for 2017 are based on preliminary data or on the last year for which data were available. For longer time series and data prior to 2016, see UNCTADstat data centre, available at http://unctadstat.unctad.org/wds/TableViewer/tableView.aspx?ReportId=32363.

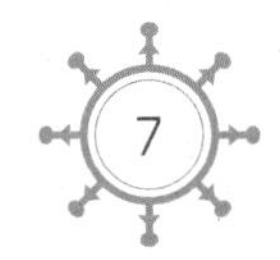


Historically, developing countries have been the main suppliers of high-volume, low-value raw materials; this has, however, changed over the years. As shown in figure 1.2, developing countries have emerged as prominent world exporters and importers. A milestone was reached in 2014 when developing countries' share of goods unloaded (imports), surpassed, for the first time, the group's share of goods loaded (exports). This shift underscores the strategic importance of developing countries as the main driver of global seaborne trade, as well as their growing participation in global value chains.

In 2004, UNCTAD noted that a new geography of trade was materializing and reshaping the global economic landscape. This new geography emphasized the growing role for the developing countries or the global South (Horner, 2016). The share of imports sourced from other developing countries increased from 37.5 per cent in 1995 to 57 per cent in 2016 (UNCTAD, 2018b).

However, participation in global value chains does not tell the whole story, as participation in these processes is not truly global but rather regional and more specifically, East Asian. Far from being a homogenous group, developing countries are not all equal when it comes to regional integration and participation in global manufacturing.

While the participation of developing countries, notably those of East Asia, in global value chains may have played a part in increasing their contribution to global goods unloaded, observed deceleration over recent years in vertical specialization suggests that factors other than participation in global value chains may also be driving growth in developing countries' seaborne imports. Overall decline in the vertical specialization process is evident when considering trade in intermediate goods. The share of intermediate imports of China as a proportion of its exports of manufacturing goods – a measure of the reliance of the manufacturing sector on imported inputs – has declined consistently over the last decade, from almost 60 per cent in 2002 to less than 40 per cent in 2014 (UNCTAD, 2016). The share of the value chain created by production abroad as a percentage of global exports is estimated to have gradually diminished since 2011, suggesting some deceleration in globalization (Berenberg and Hamburg Institute of International Economics, 2018). UNCTAD (2018c) finds that the rate of expansion of international production is slowing down, and international production and cross-border exchanges of factors of production are gradually shifting from tangible to intangible forms.

In this context, other potential factors that may be driving the continued structural shift in world seaborne trade include growth in South–South trade that is not necessarily generated by global value chains and manufacturing processes. Another potential driver is the growing consumption requirements of a fast-growing middle class in developing regions.

**Figure 1.2** **Participation of developing countries in seaborne trade, selected years**
(Percentage share in world tonnage)

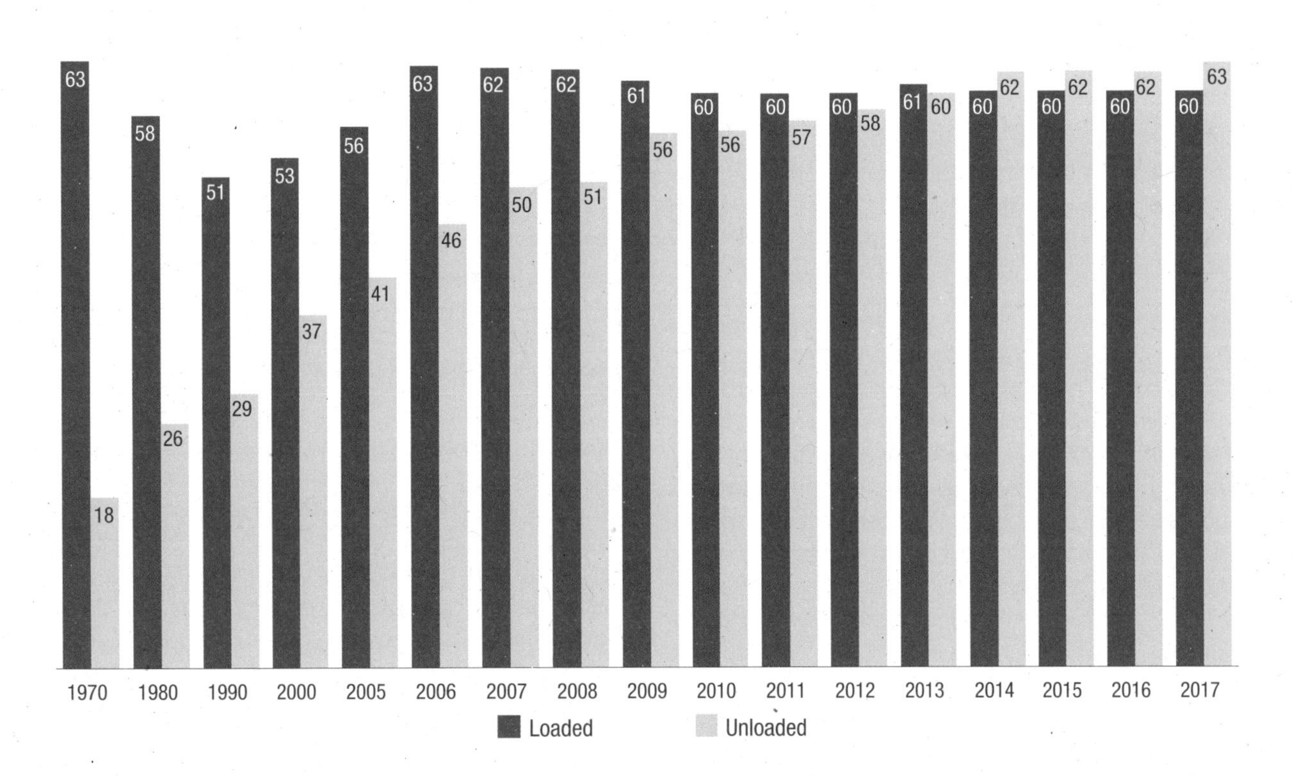


*Source:* UNCTAD secretariat calculations, based on the *Review of Maritime Transport*, various issues, and table 1.4 of this report.

**Figure 1.3 World seaborne trade, by region, 2017**
(Percentage share in world tonnage )

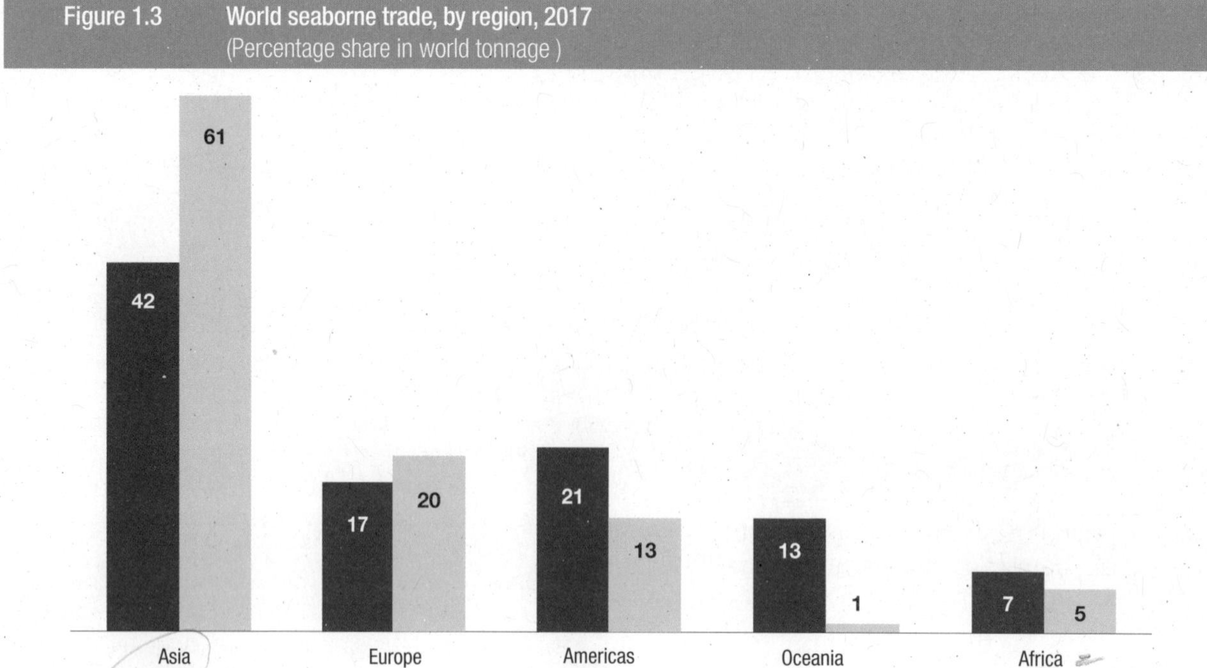


*Source:* UNCTAD secretariat calculations, based on data supplied by reporting countries and as published on government and port industry websites, and by specialist sources.
*Note:* Estimated figures are based on preliminary data or on the last year for which data were available.

Figure 1.3 highlights the leading influence of Asia, as 41 per cent of world maritime trade in 2017 originated in Asia and 61 per cent was destined to the region. Other regions, ranked in descending order, were Europe, the Americas, Oceania and Africa.

## 3. Factors contributing to more ton-miles in 2017

Seaborne trade measured in ton-miles to reflect distances travelled and the employment of ship capacity increased by 5 per cent in 2017, up from 3.41 per cent in 2016. Overall ton-miles generated by seaborne trade in 2017 amounted to an estimated 58,098 billion tons (figure 1.4). Much of the growth was driven by crude oil and coal shipments, which have greatly benefited the shipping industry, given the growth in volumes and distances. Crude oil trade contributed 17.5 per cent to ton-mile growth while major dry bulks contributed nearly one third. Together, minor bulks and other dry cargo accounted for 17.7 per cent of ton-mile growth, while containerized shipments contributed 17.4 per cent. The contributions of gas and petroleum products were much smaller.

Tanker trade ton-miles, including crude oil and refined petroleum products, rose by 4.4 per cent, and major dry bulks and containerized trade ton-miles increased by 5.5 per cent and 5.6 per cent, respectively. Minor bulks ton-miles increased by 4.5 per cent, reflecting to some extent the positive contribution of the long-distance Guinea–China bauxite trade.

Growth in tanker ton-miles was supported by firm import demand in China, as well as its oil supply diversification strategy, which is aimed at reducing the country's reliance on Western Asian crude oil. As China has been sourcing more crude oil from the Atlantic basin (countries such as Angola, Brazil, Canada, Nigeria and the United States), the number of global crude oil ton-miles has been rising. Distances travelled by crude oil trade averaged 5,047.9 nautical miles in 2017, compared with 4,941.1 nautical miles in 2016.

Growth in oil product ton-miles increased at a slower pace compared with the previous year, owing to short average sailing distances. The lifting of the United States restrictions on crude oil exports in 2015, combined with increased demand from Asia and Europe have caused crude oil seaborne exports from the United States to surpass the country's seaborne exports of oil products in terms of billion ton-miles. In 2017, global liquefied natural gas ton-miles increased by 11.6 per cent. Growing exports of liquefied natural gas from the United States underpinned growth in the average haul of imports of this commodity to China.

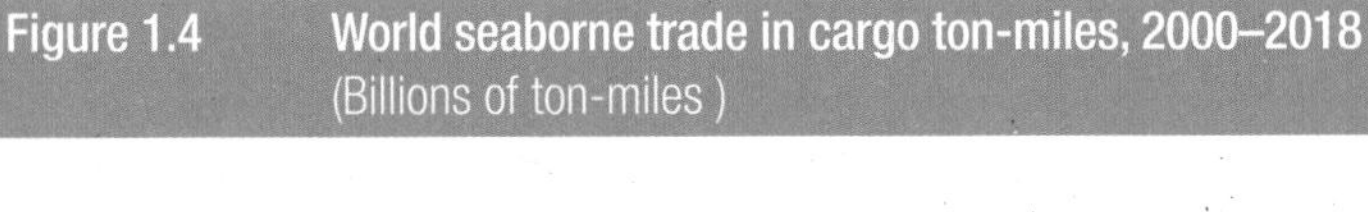

**Figure 1.4 World seaborne trade in cargo ton-miles, 2000–2018**
(Billions of ton-miles )

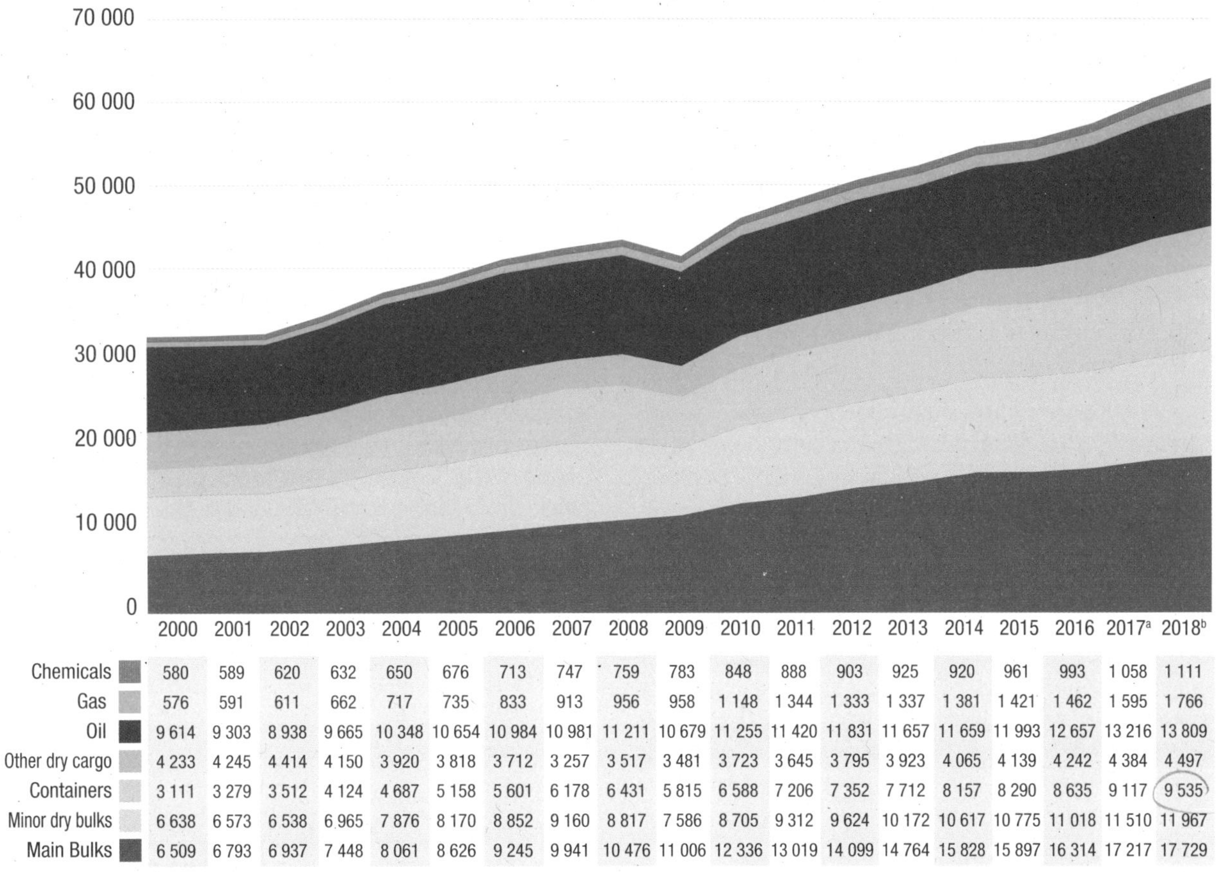


| | 2000 | 2001 | 2002 | 2003 | 2004 | 2005 | 2006 | 2007 | 2008 | 2009 | 2010 | 2011 | 2012 | 2013 | 2014 | 2015 | 2016 | 2017[a] | 2018[b] |
|---|---|---|---|---|---|---|---|---|---|---|---|---|---|---|---|---|---|---|---|
| Chemicals | 580 | 589 | 620 | 632 | 650 | 676 | 713 | 747 | 759 | 783 | 848 | 888 | 903 | 925 | 920 | 961 | 993 | 1 058 | 1 111 |
| Gas | 576 | 591 | 611 | 662 | 717 | 735 | 833 | 913 | 956 | 958 | 1 148 | 1 344 | 1 333 | 1 337 | 1 381 | 1 421 | 1 462 | 1 595 | 1 766 |
| Oil | 9 614 | 9 303 | 8 938 | 9 665 | 10 348 | 10 654 | 10 984 | 10 981 | 11 211 | 10 679 | 11 255 | 11 420 | 11 831 | 11 657 | 11 659 | 11 993 | 12 657 | 13 216 | 13 809 |
| Other dry cargo | 4 233 | 4 245 | 4 414 | 4 150 | 3 920 | 3 818 | 3 712 | 3 257 | 3 517 | 3 481 | 3 723 | 3 645 | 3 795 | 3 923 | 4 065 | 4 139 | 4 242 | 4 384 | 4 497 |
| Containers | 3 111 | 3 279 | 3 512 | 4 124 | 4 687 | 5 158 | 5 601 | 6 178 | 6 431 | 5 815 | 6 588 | 7 206 | 7 352 | 7 712 | 8 157 | 8 290 | 8 635 | 9 117 | 9 535 |
| Minor dry bulks | 6 638 | 6 573 | 6 538 | 6 965 | 7 876 | 8 170 | 8 852 | 9 160 | 8 817 | 7 586 | 8 705 | 9 312 | 9 624 | 10 172 | 10 617 | 10 775 | 11 018 | 11 510 | 11 967 |
| Main Bulks | 6 509 | 6 793 | 6 937 | 7 448 | 8 061 | 8 626 | 9 245 | 9 941 | 10 476 | 11 006 | 12 336 | 13 019 | 14 099 | 14 764 | 15 828 | 15 897 | 16 314 | 17 217 | 17 729 |

*Source:* UNCTAD secretariat calculations, based on data from Clarksons Research, 2018a.
[a] Estimated.
[b] Forecast.w

## B. WORLD SEABORNE TRADE BY CARGO TYPE

The overall positive operating environment in 2017 has benefited global demand for shipping services. However, a closer look at seaborne trade by commodity type provides a clearer picture as to the extent of the recovery.

### 1. Tanker shipments

The year 2017 witnessed the geographical dispersion of oil trade, as oil trade patterns became less concentrated on usual suppliers from Western Asia and benefited from increased trade flows from the Atlantic basin to East Asia. These trends have supported and boosted long-haul tanker trade and tanker demand. Crude oil seaborne trade expanded at a slower pace – 2.4 per cent in 2017 – compared with stronger growth – 4 per cent – in 2016 (table 1.5).

UNCTAD estimates world crude oil trade in 2017 at 1.87 billion tons, supported by increasing exports from the United States, rising global refining activity – especially in Asia – declining oil inventories and steady crude oil shipments from Western Asia. Crude oil trade benefited from the growing export volumes originating in the Atlantic basin and destined to Asia, most notably China, where rising demand from independent refiners and growing state refinery capacity boosted demand growth. An overview of global players in the oil and gas sector is presented in table 1.6.

**Table 1.5 Oil and gas trade 2016–2017**
(Million tons and percentage annual change)

| | 2016 | 2017 | Percentage change 2016–2017 |
|---|---|---|---|
| Crude oil | 1 831.4 | 1 874.9 | 2.4 |
| Other tanker trade | 1 223.7 | 1 271.2 | 3.9 |
| *of which* | | | |
| Liquefied natural gas | 268.1 | 293.8 | 9.6 |
| Liquefied petroleum gas | 87.5 | 89.3 | 2.0 |
| **Total tanker trade** | **3 055.1** | **3 146.1** | **3.0** |

*Source:* UNCTAD secretariat calculations, based on table 1.4 of this report.
*Note:* Liquefied natural gas and liquefied petroleum gas figures are derived from Clarksons Research, 2018b.

In view of the two-digit growth rate recorded in 2016 and 9.1 per cent growth experienced in 2017, China is clearly emerging as a leading importer of crude oil. Its main crude oil suppliers were Angola, the Islamic Republic of Iran, Iraq, Oman, the Russian Federation, Saudi Arabia and the Bolivarian Republic of Venezuela.

Exports from member countries of the Organization of the Petroleum Exporting Countries, especially from Western Asia, were hampered by the production cuts agreed in November 2016 and the decline in shipments from the Bolivarian Republic of Venezuela. These trends were, however, offset by growing shipments from the United States, reflecting the rapid growth in its shale oil output, as well as a recovery in exports from Libya and Nigeria.

Together, refined petroleum products and gas volumes increased by 3.9 per cent in 2017; growth in petroleum products was supported by rising demand in developing America and growing intra-Asian trade. However, elevated global inventory and stocks undermined arbitrage opportunities for some products and hindered growth during the year. At the same time, drawdowns on inventories weighed on the import demand in some regions, including Europe (Clarksons Research, 2018a).

**Table 1.6 Major producers and consumers of oil and natural gas, 2017**
(World market share, in percentage)

| World oil production | | World oil consumption | |
|---|---|---|---|
| Western Asia | 34 | Asia and the Pacific | 35 |
| North America | 19 | North America | 23 |
| Transition economies | 15 | Europe | 15 |
| Developing America | 10 | Western Asia | 10 |
| Africa | 9 | Developing America | 9 |
| Asia and the Pacific | 9 | Transition economies | 4 |
| Europe | 4 | Africa | 4 |
| **Oil refinery capacities** | | **Oil refinery throughput** | |
| Asia and the Pacific | 34 | Asia and the Pacific | 35 |
| North America | 21 | North America | 22 |
| Europe | 15 | Europe | 16 |
| Western Asia | 10 | Western Asia | 10 |
| Transition economies | 9 | Transition economies | 8 |
| Developing America | 8 | Developing America | 6 |
| Africa | 3 | Africa | 3 |
| **World natural gas production** | | **World natural gas consumption** | |
| North America | 25 | North America | 23 |
| Transition economies | 22 | Asia and the Pacific | 21 |
| Western Asia | 18 | Transition economies | 16 |
| Asia and the Pacific | 17 | Western Asia | 15 |
| Europe | 7 | Europe | 14 |
| Developing America | 6 | Developing America | 7 |
| Africa | 5 | Africa | 4 |

*Source:* UNCTAD secretariat calculations, based on data from British Petroleum, 2018.

*Notes:* Oil includes crude oil, shale oil, oil sands and natural gas liquids. The term excludes liquid fuels from other sources such as biomass and coal derivatives.

On the supply side, higher levels of refinery throughput lifted export volumes from Europe and Asia, including Western Asia and China. The United States contributed to export growth, and shipments of oil products expanded by 9.5 per cent (Clarksons Research, 2018b). United States exports to developing America partly benefited from the continued decline in refinery activity in Brazil, Mexico and the Bolivarian Republic of Venezuela.

Growing domestic refinery capacity has increasingly positioned China as a significant exporter of oil products, with its export volumes more than doubling between 2013 and 2016 (Clarksons Research, 2018c). Although less impressive than the 2016 surge of more than 50 per cent, exports from China increased by 6.3 per cent in 2017, driven by the ongoing oversupply of oil products in that country. The deceleration observed in 2017 partly reflects its growing domestic consumption requirements.

## 2. Factors supporting trade in gas and refined petroleum products

Shipments of liquefied natural gas totalled 293.8 million tons in 2017, following a 9.6 per cent increase over the previous year (table 1.5) (Clarksons Research, 2018b). Increased demand, the highest in six years, originated mostly in Asia, where energy policy shifts are under way. Imports of the commodity to China increased by 47.3 per cent in 2017, owing to weather conditions and stronger demand. The country's demand for liquefied natural gas was partly supported by the growing importance of the environmental agenda. Further, the continued expansion of liquefied natural gas regasification capacity in China highlights the potential for further expansion in imports of the commodity.

Key exporters included Qatar, which remained the largest supplier of liquefied natural gas. Other exporters were Australia, the Russian Federation and the United States. Much of the growth was underpinned by increased exports from Australia to Asia, although long-haul trade from the United States to Asia was on the rise. Increased production from liquefied natural gas projects commissioned in 2016 and the start of operations at liquefication facilities in Australia, the Russian Federation and the United States, boosted export volumes of the commodity. During the year, the world's first floating liquefied natural gas facility started operations in Malaysia (Barry Rogliano Salles, 2018), and one project received approval in Mozambique, a major development, given the rise of the country as a producer of liquefied natural gas.

Shipments of liquefied petroleum gas expanded at a slower pace (2.0 per cent) in 2017, down from 11.2 per cent in 2016 (Clarksons Research, 2018b). The main factors restricting growth included a decline in Western

Asian exports, which was offset somewhat by growing exports from the United States. Demand for imports in China was key, with import volumes expanding by 14.7 per cent. This pace is, however, less than half of that in 2016 (34.4 per cent), reflecting the end of the recent wave of propane dehydrogenation plant expansions (Danish Ship Finance, 2017). Imports of liquefied petroleum gas to India increased in 2017, supported by a subsidy programme of the Government promoting households' switch to cleaner fuels. In contrast, imports of the commodity to Europe declined, owing in part to competition from ethane. With regard to chemicals, volumes also increased following the growing demand for imports in Asia, a rebound in palm oil trade after El Niño in 2016 and growth in United States exports.

## 3. Dry-cargo trades: The mainstay of seaborne trade in 2017

### *Dry bulk shipments: Major and minor dry bulks*

Following a limited expansion in 2015–2016, global dry bulk trade[1] grew by about 4 per cent in 2017, bringing total volumes to 5.1 billion tons (table 1.7). A sharp increase in iron ore imports to China, a rebound in global coal trade and improved growth in minor bulk trades supported the expansion. Overall, strong import demand in China remained the main factor behind growth in global dry bulk trade. An overview of global players in the dry bulk commodities trade sector is presented in table 1.8.

#### *Iron ore*

Iron ore imports to China increased by 5 per cent in 2017, bringing total volumes to nearly 1.1 billion tons. With a market share of more than 70 per cent, China remains the main source of global iron ore demand. A rise in steel production and the closure of more than 100 million tons per annum of outdated steelmaking capacity in 2016–2017 boosted the country's demand for imports. Further, the increased use of higher grade imported iron ore displaced domestic supplies. The leading iron ore exporters were Australia, Brazil and South Africa; Australia and Brazil supplied over 85 per cent of the demand for imports in China. Nevertheless, Australia is by far the largest exporter, supplying nearly two thirds of iron ore requirements in China. The country imports 21 per cent of its iron ore requirements from Brazil, which benefits the dry bulk shipping industry through long distances. South Africa generates 4 per cent of all iron ore imports to China. Other suppliers, such as India, the Islamic Republic of Iran and Sierra Leone, have also increased their exports to China.

**Table 1.7 Dry bulk trade 2016–2017** (Million tons and percentage annual change)

| | 2016 | 2017 | Percentage change 2016–2017 |
|---|---|---|---|
| *Main bulks* | 3 040.9 | 3 196.3 | 5.1 |
| *of which:* | | | |
| Iron ore | 1 418.1 | 1 472.7 | 3.9 |
| Coal | 1 141.9 | 1 208.5 | 5.8 |
| Grain | 480.9 | 515.1 | 7.1 |
| *Minor bulks* | 1 874.6 | 1 916.5 | 2.2 |
| *of which:* | | | |
| Steel products | 406.0 | 390.0 | -3.9 |
| Forest products | 354.6 | 363.6 | 2.5 |
| **Total dry bulks** | **4 915.5** | **5 112.8** | **4.0** |

*Source:* UNCTAD secretariat calculations, based on Clarksons Research, 2018a.

#### *Coal*

Global coal trade resumed growth in 2017, increasing by 5.8 per cent following a limited expansion in 2016 and a significant decline in 2015. Higher import demand in China, the Republic of Korea and a number of South-East Asian countries supported the volume increase. Coal imports to China continued to provide strong support for dry bulk shipping demand. China, India, Japan, Malaysia, and the Republic of Korea are major importers of coal, while Australia and Indonesia are major exporters of the commodity. Growing coal exports from the United States to China are benefiting dry bulk shipping. One factor is the uncertainty over the Indian coal trade. On the one hand, India plans to increase domestic production, which may alter the balance between locally sourced and imported coal. On the other hand, growing demand from the steel sector in India may boost seaborne imports of coking coal (Barry Rogliano Salles, 2018).

#### *Grain*

Global grain trade, including wheat, coarse grains and soybeans, reached 515.1 million tons in 2017, a 7.1 per cent increase over 2016. Exports are dominated by a few countries, notably the United States; importers tend to be regionally diverse.

As in other dry bulk trades, Asia was a driving force of growth, albeit not the only one. In 2017, grain trade was underpinned by a 14.7 per cent increase in soybean imports to China and growing exports from Brazil and the United States. China dominates the soybean trade and accounted for nearly two thirds of the global soybean import demand in 2017. Outside Asia and the European Union, some lesser consuming regions, such as Africa and Western Asia, also contributed to such growth.

Tariffs by the United States on certain goods imported from China, including steel and aluminium, and retaliation by China, may lead to restricting soybean import from the United States. China is the world's largest consumer and importer of uncrushed soybeans. However, it may decide to replace imports from the United States and source its soybean requirements from alternative suppliers such as Brazil. While trade restrictions generally portend ominous consequences for shipping, a shift in suppliers and routes in this context may have an unintended positive effect on ton-miles generated.

### *Minor bulks*

Growing manufacturing activity and construction demand supported a 2.2 per cent increase in minor bulks commodity trade. Rising demand for commodities such as bauxite, scrap and nickel ore pushed volumes to 1.9 billion tons. However, the large drop (less 30.8 per cent) in exports of steel products from China due to reforms in the country's steel sector undermined the expansion to some extent. Bauxite shipments expanded by 19.5 per cent, accounting for 13 per cent of minor dry bulks commodities trade in 2017. The continued rise in Chinese aluminium production and the availability of bauxite ore, following years of export disruptions, led to an expansion in bauxite trade. While China dominates the import side with a market share of more than two thirds, key players on the supply side are more varied and include Australia, Brazil, Guinea and India. Nickel ore trade rose by 7.6 per cent, highlighted in particular by increased growth in nickel ore shipments from Indonesia, following its decision to relax its export ban on unprocessed ores.

**Table 1.8 Major dry bulks and steel: Producers, users, exporters and importers, 2017**
(World market shares, in percentage)

| Steel producers | | Steel users | |
|---|---|---|---|
| China | 49 | China | 46 |
| Japan | 6 | United States | 6 |
| India | 6 | India | 5 |
| United States | 5 | Japan | 4 |
| Russian Federation | 4 | Republic of Korea | 4 |
| Republic of Korea | 4 | Germany | 3 |
| Germany | 3 | Russian Federation | 3 |
| Turkey | 2 | Turkey | 2 |
| Brazil | 2 | Mexico | 2 |
| Other | 19 | Other | 25 |
| **Iron ore exporters** | | **Iron ore importers** | |
| Australia | 56 | China | 72 |
| Brazil | 26 | Japan | 9 |
| South Africa | 4 | Europe | 8 |
| Canada | 3 | Republic of Korea | 5 |
| India | 2 | Other | 6 |
| Other | 9 | | |
| **Coal exporters** | | **Coal importers** | |
| Indonesia | 32 | China | 18 |
| Australia | 30 | India | 17 |
| Colombia | 7 | Japan | 15 |
| United States | 7 | Europen Union | 13 |
| South Africa | 7 | Republic of Korea | 12 |
| Canada | 2 | Taiwan Province of China | 6 |
| Other | 15 | Malaysia | 3 |
| | | Other | 16 |
| **Grain exporters** | | **Grain importers** | |
| United States | 25 | East and South Asia | 34 |
| Russian Federation | 23 | Africa | 21 |
| Ukraine | 15 | Developing America | 20 |
| Argentina | 11 | Western Asia | 16 |
| Europena Union | 9 | Europe | 7 |
| Australia | 8 | Transition economies | 2 |
| Canada | 7 | | |
| Other | 2 | | |

*Source:* UNCTAD secretariat calculations, based on data from Clarksons Research, 2018d and World Steel Association, 2018a, 2018b.

### *Other dry cargo: Containerized trade*

Following the difficult years of 2015 and 2016 when containerized trade grew modestly at 1.1 per cent and 3.1 per cent, respectively, container market conditions improved in 2017, and strong growth in volumes was recorded across all routes. World containerized trade volumes expanded by a strong 6.4 per cent in 2017, the fastest rate since 2011. Global volumes reached 148 million TEUs (figure 1.5), supported by various positive trends.

The modest global recovery was central to the rise in containerized volumes. In addition, factors such as a recession in Brazil and the Russian Federation, increased consumption requirements in the United States, improved commodity prices, strong import demand from China and the rapid growth of intra-Asian trade reflecting the effect of regional integration and participation in global value chains, contributed to the recovery.

Trade growth strengthened on the major East–West trade lanes, namely Asia–Europe, the Trans-Pacific and transatlantic routes (table 1.9 and figure 1.6). Volumes on the Trans-Pacific route (eastbound and westbound) increased by 4.7 per cent, while volumes on the East Asia–North America route (eastbound and westbound) increased by 7.1 per cent. Overall, the Trans-Pacific trade lane remained the busiest, with total volumes reaching 27.6 million TEUs, followed by 24.8 million TEUs on the Asia–Europe route and 8.1 million TEUs on the transatlantic route.

Growth accelerated across non-mainlane routes (table 1.10). Robust growth (6.5 per cent) on the North–South trade route reflected improvements in the commodity price environment and the higher import demand of oil- and commodity-exporting countries. Supported by positive economic trends in China, economic growth in emerging Asian economies, as well as regional integration and global value chains, volumes on the intra-Asian routes picked up, expanding by 6.7 per cent. Containerized trade on the non-mainlane East–West routes grew by an estimated 4.0 per cent, with varied performances across individual routes; key factors were faster growth on routes within and outside the Indian subcontinent and slower growth on routes within and outside Western Asia.

**Figure 1.5 Global containerized trade, 1996–2018**
(Million 20-foot equivalent units and percentage annual change)

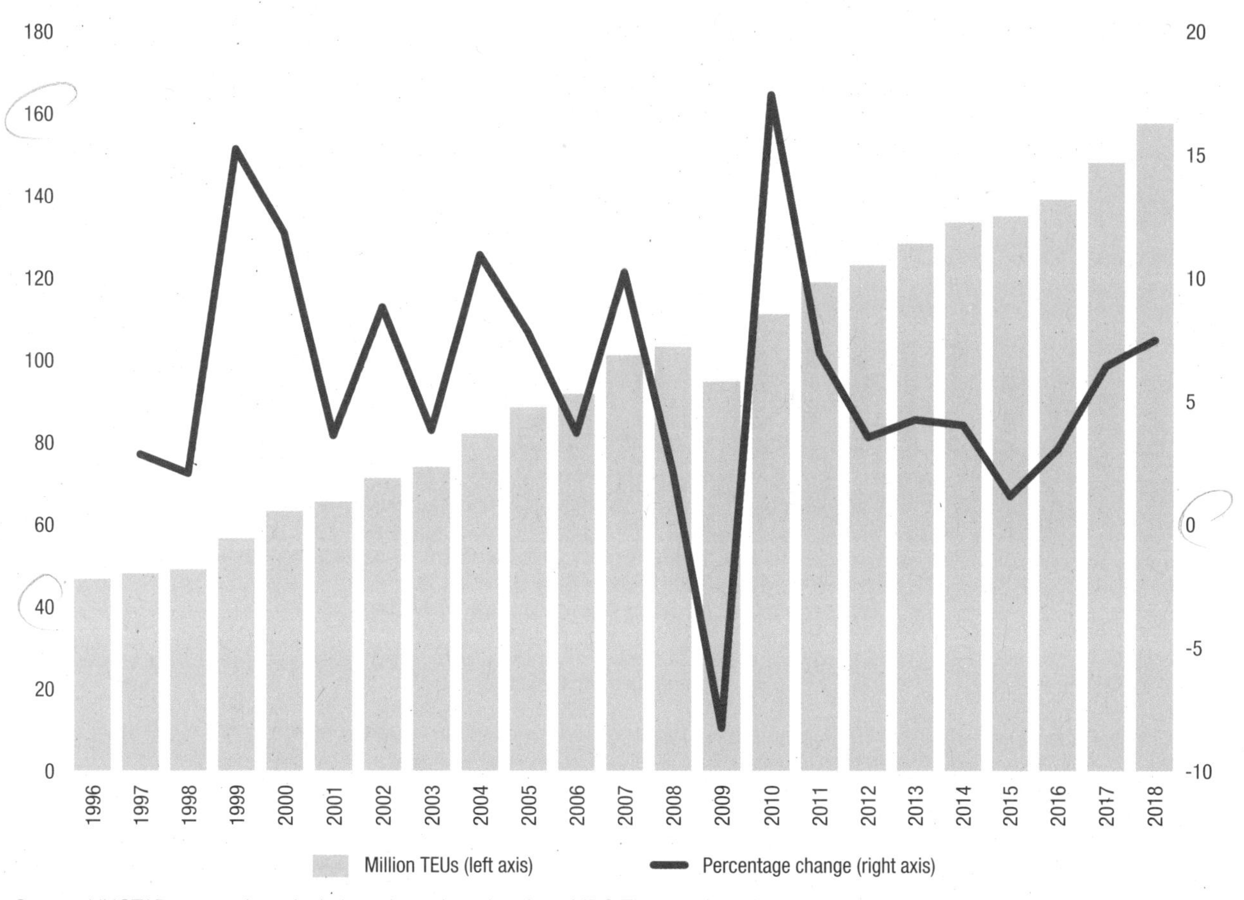


*Source:* UNCTAD secretariat calculations, based on data from MDS Transmodal, 2018.
*Note:* Data for 2018 are projected figures.

**Table 1.9 Containerized trade on major East–West trade routes, 2014–2018**
(Million 20-foot equivalents and percentage annual change)

| | Trans-Pacific | | Asia–Europe | | Transatlantic | |
|---|---|---|---|---|---|---|
| | Eastbound | Westbound | Eastbound | Westbound | Eastbound | Westbound |
| | East Asia–North America | North America–East Asia | Northern Europe and Mediterranean to East Asia | East Asia to Northern Europe and Mediterranean | North America to Northern Europe and Mediterranean | Northern Europe and Mediterranean to North America |
| 2014 | 15.8 | 7.4 | 6.8 | 15.2 | 2.8 | 3.9 |
| 2015 | 16.8 | 7.2 | 6.8 | 14.9 | 2.7 | 4.1 |
| 2016 | 17.7 | 7.7 | 7.1 | 15.3 | 2.7 | 4.2 |
| 2017 | 18.7 | 7.9 | 7.6 | 16.4 | 3.0 | 4.6 |
| 2018[a] | 19.5 | 8.1 | 7.8 | 16.9 | 3.2 | 4.9 |
| **Percentage annual change** | | | | | | |
| 2014–2015 | 6.6 | -2.9 | 0.2 | -2.3 | -2.4 | 5.6 |
| 2015–2016 | 5.4 | 7.3 | 3.8 | 2.7 | 0.5 | 2.8 |
| 2016–2017 | 5.6 | 2.1 | 6.9 | 7.1 | 8.0 | 8.3 |
| 2017–2018[a] | 4.1 | 3.0 | 3.2 | 3.3 | 7.3 | 7.1 |

*Source:* UNCTAD secretariat calculations, based on MDS Transmodal, 2018.
[a] Forecast.

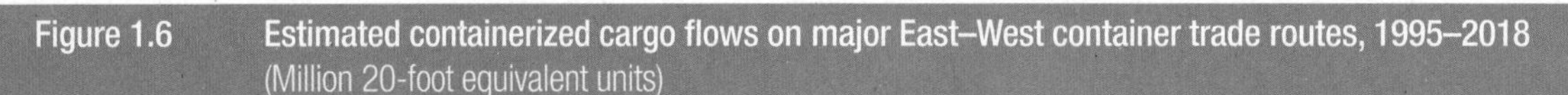

**Figure 1.6** **Estimated containerized cargo flows on major East–West container trade routes, 1995–2018**
(Million 20-foot equivalent units)

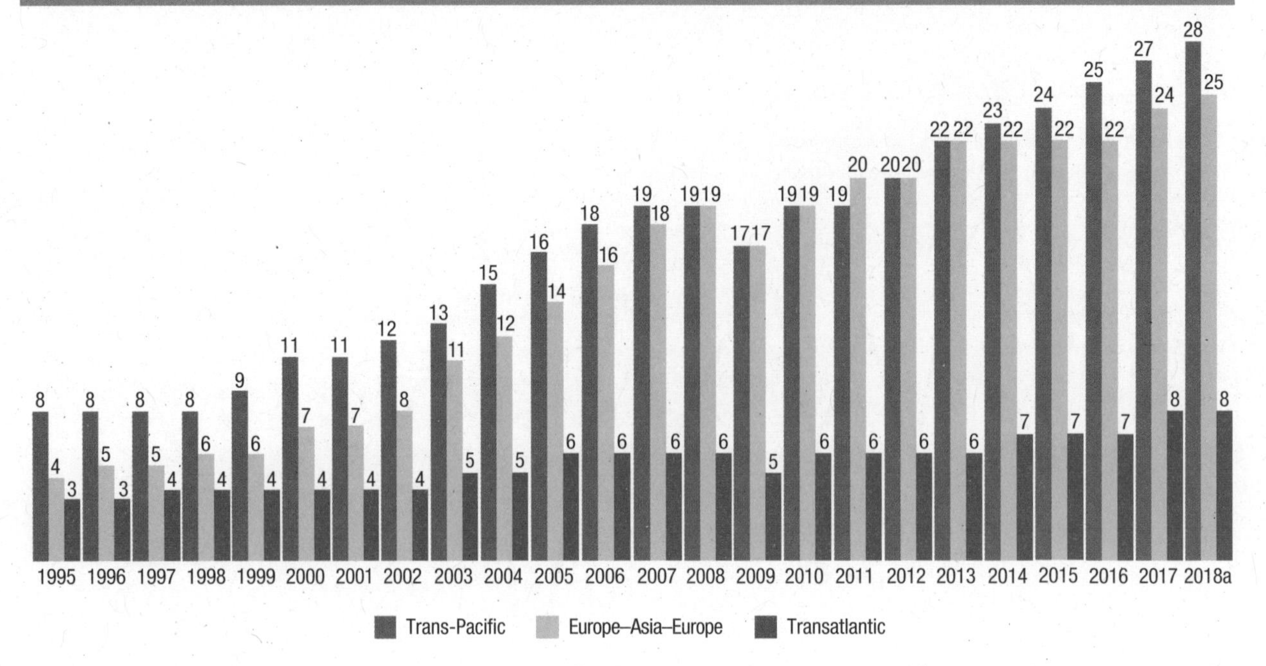


*Source:* UNCTAD secretariat calculations, based on Economic Commission for Latin America and the Caribbean, 2010. Figures from 2009 onward are derived from data provided by MDS Transmodal and Clarksons Research.
[a] Forecast.

**Table 1.10** **Containerized trade on non-mainlane routes, 2016–2018**
(Million 20-foot equivalents and annual percentage change)

| Intraregional | | Intra-Asian | Non-mainlane East–West | North–South |
|---|---|---|---|---|
| Percentage annual change | | | | |
| 2016 | 5.0 | 5.6 | 4.9 | 1.9 |
| 2017 | 6.3 | 6.7 | 4.0 | 6.5 |
| 2018[a] | 6.1 | 6.8 | 5.2 | 6.4 |

*Source:* UNCTAD secretariat calculations, based on data from Clarksons Research, 2018e.
[a] Forecast.

Positive trends in the containerized trade market unfolded against the backdrop of continued market consolidation; alliance reshuffling; ordering of larger ships, with capacities likely to stabilize at close to 20,000–22,000 TEUs; as well as a growing momentum surrounding e-commerce and digitalization. Together these factors are reshaping the containerized trade and liner shipping landscape and raising new challenges and opportunities for the sector.

The rise of mega alliances is likely to reinforce the commoditization of container transportation services, as they tend to limit liner shipping service or product differentiation (McKinsey and Company, 2017a). This means that lines would be unable to differentiate themselves and to compete based on service. As a member of an alliance, a shipping line may not be able to offer faster and more reliable services than its alliance partners. For shippers, the commoditization of services would also be an unfavourable development, as it limits their ability to obtain greater transparency and reliability, as well as the right services. This is because shippers do not know which ship or operator is handling their cargo in an alliance arrangement. Overall, it seems that alliances help to expand the service range available but tend to heighten operational complexities and detract from transparency along the logistics chain (see chapters 2 and 3).

### *Electronic commerce*

The rapid expansion of e-commerce is of direct relevance to the container shipping market, given the related implications for consumption patterns, retail models, distribution networks, and transport and logistics. UNCTAD estimates global e-commerce at almost $26 trillion in 2016 (UNCTAD, 2018d). Cross-border e-commerce is particularly relevant to shipping and accounts for a relatively smaller share of total e-commerce in general and business-to-consumer sales, in particular. According to UNCTAD, such cross-border transactions were worth about $189 billion in 2015. Dwarfed by the size of domestic business-to-consumer e-commerce, cross-border sales in that year accounted for 6.5 per cent of total business-to-consumer e-commerce (UNCTAD, 2017a). Nevertheless, business-to-consumer e-commerce, including cross-border transactions, is growing rapidly, and Asia is becoming a major growth area. While data on e-commerce trends in developing countries are difficult to obtain, cross-border e-commerce in China was said to account for up to 20 per cent of total import and

export trading volumes (JOC.com, 2017). Elsewhere in the region, the size of e-commerce-related business is much smaller, but is characterized by rapid growth. In India, e-commerce sales were estimated at around $40 billion in 2016, up from $4 billion in 2009, while in Indonesia, the market was worth about $6 billion in 2016. By 2020, 45 per cent of online shoppers are expected to buy goods from other countries. This would represent a fourfold increase in the value of cross-border sales since 2014 (Colliers International, 2017).

Shipping, like other modes of transport, is also part of the e-commerce supply chain. However, the extent to which container shipping is able to benefit from e-commerce trade flows and capture some of the associated gains remains unclear in view of the relatively small share of cross-border business-to-consumer e-commerce flows and the participation of alternative modes of transport. The speed of air transport favourably positions aviation as a better fit for e-commerce trade, notably for high-value and time-sensitive cargo. Rail transport could also gain market share as illustrated by developments in the China–Europe rail connections and the example offered by the China–Germany service advertised on the Alibaba portal (Colliers International, 2017). Nevertheless, ocean shipping is expected to contribute to e-commerce trade and benefit from the transport of other goods and products that rests on the building of inventories near consumption markets.

For shipping to tap the trade potential arising from e-commerce, operators need to adapt, leverage technology for greater efficiencies and design integrated supply chain solutions that are e-commerce-friendly. Adaptation and planning for change is critical for shipping to remain a relevant market player. In this respect, concerns have recently been raised over the potential for e-retailers to displace traditional players such as liner shipping operators. While these concerns have generally been downplayed, shipping lines recognize the potential risks and seem to be adapting their business models to account for these emerging trends, including by leveraging technology and digitalization to ensure efficiency gains and capture market share. An example is the new global integrator strategy pursued by Maersk to drive down costs, improve reliability, enhance responsiveness and forge a better link with customers (Maersk, 2018).

#### *Digitalization*

Today, the shipping industry is cautiously embracing relevant technologies arising from digitalization. More and more, carriers and freight forwarders alike are taking measures to digitalize internal processes, develop integrated information technology infrastructures and offer real-time transparency on shipments. Digital start-ups such as Xeneta, Flexport and Kontainers are being launched (McKinsey and Company, 2017b). These solutions aim to provide user-friendly online interfaces for shippers, while facilitating processes and enhancing transparency. Recent developments relating to blockchain technology aimed at facilitating seaborne trade are also important (see chapter 5). Some argue that the technology could save $300 in customs clearance costs for each consignment and that it could potentially generate $5.4 million in savings on each shipment associated with a ship that has a capacity of 18,000 TEUs (Marine and Offshore Technology, 2017).

Other technologies of relevance to seaborne trade include robotics, artificial intelligence and additive manufacturing or three-dimensional printing. Robotics have some implications for production localization by enabling zero-labour factories (Danish Ship Finance, 2017). According to UNCTAD research however, robot use in low-wage labour-intensive manufacturing has remained low (UNCTAD, 2017b).

Three-dimensional printing and robotics may facilitate regionalized manufacturing and lead to some reshoring by displacing low-cost labour. While three-dimensional printing, in particular, is not expected to cause a massive relocalization pattern, it may have an incremental impact and affect specific niche markets. In time, this technology may lead to less raw materials being used in manufacturing. Until it becomes widespread and cost-effective, for now the impact of three-dimensional printing is expected to be marginal – existing estimates suggest that TEU volumes will drop by less than 1 per cent by 2035 (JOC.com, 2017).

## C. OUTLOOK AND POLICY CONSIDERATIONS

### 1. World seaborne trade projections: 2018–2023

Global seaborne trade is doing well, helped by the upswing in the world economy. Prospects for the short and medium term are positive overall – global GDP is expected to grow by more than 3.0 per cent over the 2018–2023 period (International Monetary Fund, 2018), and merchandise trade volumes are set to rise by 4.4 per cent in 2018 and 4 per cent in 2019 (World Trade Organization, 2018). In line with projected economic growth and based on the income elasticity of seaborne trade estimated for the 2000–2017 period, UNCTAD expects world seaborne trade volumes to expand by 4.0 per cent in 2018. According to UNCTAD projections, world seaborne trade will expand at a compound annual growth of 3.8 per cent during that period, based on calculated elasticities and the latest figures of GDP growth forecast by the International Monetary Fund for 2018–2023. Overall, these projections are comparable with existing ones, such as those by Clarksons Research and Lloyd's List Intelligence (table 1.11). Further, they are consistent with past trends indicating that seaborne trade increased at an annual average growth rate of 3.5 per cent between 2005 and 2017 and that dry bulk commodities and containerized trades have been driving much of the growth.

Contingent on continued economic conditions in the global economy, volumes across all segments are set to expand; it is expected that containerized and dry bulk commodities trades will record the fastest growth. Tanker trade volumes should increase, although at a slightly slower pace than other cargo types. Dry bulk commodities are projected to experience a compound annual growth rate of 4.9 per cent between 2018 and 2023, while containerized shipments are expected to rise by 6 per cent, supported by positive economic trends, imports of metal ores to China and steady growth on the non-mainlane trade routes. Further, crude oil trade is forecast to grow by 1.7 per cent between 2018 and 2023, and combined petroleum products and gas volumes, by 2.6 per cent.

The positive outlook for seaborne trade could be sustained by the trade liberalization gains that may be generated by various trade policy instruments, providing they are successfully concluded and implemented. These include the Comprehensive and Progressive Agreement for Trans-Pacific Partnership, the Agreement between the European Union and Japan for an Economic Partnership, the trade and investment agreements between the European Union and Singapore,[2] the Regional Comprehensive Economic Partnership and the Agreement Establishing the African Continental Free Trade Area. The latter agreement, according to UNCTAD, could increase the value of intra-African trade by 33 per cent (UNCTAD, 2018e).

While the advantages and implications of the implementation of the Agreement Establishing the African Continental Free Trade Area with regard to seaborne trade are yet to be fully assessed, additional trade flows can be expected to benefit shipping and support seaborne trade volumes (Brookings Instituion, 2018). In this respect, one liner shipping operator reported that intra-Africa trade had picked up following the implementation of trade facilitation measures, in particular the one-stop border post concept (Southern Africa Shipping News, 2017). This points to the significant potential in Africa that could be unlocked for shipping and seaborne trade if relevant support measures and enabling conditions were to be provided.

Growing intra-Asian trade arising from a shift of low-cost manufacturing activities from China to other neighbouring East and South Asian countries could generate some additional seaborne trade flows. As China moves up the global value chain, new trading opportunities are opening up for other countries. The value of outward-oriented greenfield foreign direct investment in manufacturing in developing Asia has nearly doubled, from $26.6 billion in 2005–2010 to $50.2 billion in 2011–2016 (Asian Development

**Table 1.11 Seaborne trade development forecasts, 2017–2026**
(Percentage change)

| | Annual growth rate | Years | Seaborne trade flows | Source |
|---|---|---|---|---|
| Lloyd's List Intelligence | 3.1 | 2017–2026 | Seaborne trade | *Lloyd's List Intelligence research, 2017* |
| | 4.6 | 2017–2026 | Containerized trade | |
| | 3.6 | 2017–2026 | Dry bulk | |
| | 2.5 | 2017–2026 | Liquid bulk | |
| Clarksons Research Services | 3.4 | 2018 | Seaborne trade | *Seaborne Trade Monitor, May 2018* |
| | 5.2 | 2018 | Containerized trade | *Container Intelligence Monthly, April 2018* |
| | 2.6 | 2018 | Dry bulk | *Dry Bulk Trade Outlook, April 2018* |
| | 2.4 | 2018 | Liquid bulk | *Seaborne Trade Monitor, May 2018* |
| | 4.9 | 2019 | Containerized trade | *Container Intelligence Monthly, April 2018* |
| Drewry Maritime Research | 4.5 | 2018 | Containerized trade | *Container Forecaster, Quarter 1, 2018* |
| | 4.2 | 2019 | Containerized trade | *Container Forecaster, Quarter 1, 2018* |
| UNCTAD | 4.0 | 2018 | Seaborne trade volume | *Review of Maritime Transport 2018* |
| | 5.2 | 2018 | Dry bulk | |
| | 6.4 | 2018 | Containerized trade | |
| | 1.8 | 2018 | Crude oil | |
| | 2.8 | 2018 | Refined petroleum products and gas | |
| | 3.8 | 2018–2023 | Seaborne trade | *Review of Maritime Transport 2018* |
| | 4.9 | 2018–2023 | Dry bulk | |
| | 6.0 | 2018–2023 | Containerized trade | |
| | 1.7 | 2018–2023 | Crude oil | |
| | 2.6 | 2018–2023 | Refined petroleum products and gas | |

*Source:* UNCTAD secretariat calculations, based on own calculations and forecasts published by the indicated institutions and data providers.

Bank, 2017). Major recipients included Cambodia, India, Indonesia, Malaysia and Thailand. Unlike China, where the growing share of domestic content used in manufacturing limits growth in intermediate goods, these countries are likely to source much of the goods from external suppliers and thus generate additional trade activity.

In addition, various projects under the Belt and Road Initiative of China have the potential to generate growth and boost seaborne trade volumes through increased demand for raw materials and semi-finished and finished products. Infrastructure developments of the size of the Initiative require large amounts of construction materials in the form of dry bulk commodities, steel products, cement, heavy machinery and equipment. Improvements in connectivity through enhanced transport infrastructure, linking manufacturing industry or agriculture to global markets, could strengthen many countries' economic growth and boost trade. These developments have favourable implications for container shipping and bulk commodities trade.

However, an expanding overland route between China and Europe that has already attracted movements of high-value, time-sensitive goods – which previously would have been transported by sea – could shift some seaborne cargo from ship to rail. The pipelines built under the framework of the Belt and Road Initiative could also restrict seaborne trade growth in related trades (Hellenic Shipping News, 2017). All in all, however, the net effect of the initiative could support shipping demand, as rail transport services and pipelines are not expected to significantly displace the role of shipping in the region and along the Asia–Europe trade lane.

As noted previously, the prospects for seaborne trade are positive and may be sustained by the various upside factors. Yet caution is required, given the uncertainty arising from the confluence of geopolitical, economic and trade policy risks, and structural shifts, such as the rebalancing of the Chinese economy, slower growth of global value chains and a change in the global energy mix. How these factors will evolve and the extent to which they will support or derail the recovery in seaborne trade remains unclear. A major trade policy risk relates to the inward-looking policies and the rise of protectionism, which may reverse the trade liberalization of today. Examples include the decision of the United States to withdraw from the Trans-Pacific Partnership Agreement, to renegotiate the North American Free Trade Agreement and to re-evaluate other existing trade agreements. Such policies can produce significant setbacks for global economic and trade recovery and undermine the growth prospects of seaborne trade.

Another risk of this nature is associated with the growing trade tensions between the United States and some of its trading partners. Following the announcement by the United States in March 2018 to apply tariffs to steel and aluminium imports, the United States, within the framework of the North American Free Trade Agreement, in May proceeded to apply such tariffs to imports from the European Union. Such developments could be detrimental for global trade, depending on how major trading partners respond to the new trade restrictions.

A closer look at the specific trades and commodities that may be affected by the United States tariffs on steel and aluminium, as well as the proposed tariffs on a list of other products imported from China, indicates that importers and exporters will be facing uncertainty and disruptions relating to dry bulk shipping (for example, steel, aluminium and soybeans), as well as some proportion of the containerized trade between China and the United States. According to one observer, tariffs currently in force in those countries affect an estimated 24 million tons of seaborne trade, equivalent to some 0.2 per cent of global seaborne trade (Clarksons Research, 2018f). If proposed tariffs were to be accounted for, the impact would increase to 0.7 per cent of world seaborne trade volume. However, this could produce an unintended positive effect – an increase in soybeans ton-miles to China – if Argentinian and Brazilian soybeans were to displace soybeans from the United States.

The list of containerized goods from China that could be affected by the proposed tariffs include furniture, electrical machinery, rubber manufactures, clothing and accessories, and metal manufactures. These goods are shipped in containers from Eastern Asia to the West Coast of the United States on the Trans-Pacific route. As the China–United States trade on this route accounts for about 3 per cent of total global containerized trade, the overall impact is not likely to be disruptive. Overall, the impact may initially be limited, depending on the duration of the tariffs and the extent of the retaliatory measures by trading partners.

Other factors and potential risks for the sustained recovery of seaborne trade and its outlook include the following:

- Trade policy risks linked to the decision by the United Kingdom of Great Britain and Northern Ireland to leave the European Union and the related implications for business confidence and investment activity in Europe. Other concerns relate to the increasing number of trade disputes that have been raised at the World Trade Organization, regarding for example, Australia, Canada, China, India, Pakistan, the Republic of Korea, the Russian Federation, Ukraine, the United Arab Emirates, the United States and Viet Nam.
- Withdrawal of the United States from the Joint Comprehensive Plan of Action and the re-imposing of international sanctions on the Islamic Republic of Iran.
- Deterioration of the economic crisis in the Bolivarian Republic of Venezuela and related implications for tanker trade and other sectors.
- The gradual transition of China towards a more diversified economy and its efforts to reduce industrial overcapacity and improve air quality. Developments in that country are important for

seaborne trade prospects, given its strategic importance for shipping demand, especially dry bulk commodities trade. In view of the significant market shares of China in trade in various dry bulks commodities – for example, iron ore, bauxite, coal and nickel ore – the slightest negative shift in its import requirements can be potentially detrimental to shipping demand.

- Structural forces, including the slower pace of trade liberalization, as well as global value chain integration. As stated in the 2017 and 2016 editions of the *Review of Maritime Transport*, cyclical factors alone do not explain the decline in the ratio of trade growth to GDP growth.
- Although beneficial for sustainability objectives, the transition of the global economy towards a less fossil fuel-intensive growth model entails some uncertainty for oil, gas and coal trades. A similar concern arises in connection with trends in the circular economy. Applying circular economy principles may hold back demand for raw materials, although it would be a boon for the sustainability agenda.
- Potentially unintended negative impacts of emerging technologies such as three-dimensional printing and robotics may cancel out the positive gains for maritime trade.

## 2. Policy considerations

UNCTAD projections are pointing to continued growth in world seaborne trade, which hinges on continued growth in GDP. At the same time, upside and downside risks to the outlook are manifold and include rising trade tensions on the downside and digitalization on the upside. Further, new factors such as digitalization, e-commerce and the Belt and Road Initiative are increasingly unfolding. Depending on their extent and the pace at which they evolve, they may alter the face of global shipping and redefine seaborne trade flows and patterns.

In this context, it is increasingly acknowledged that the value of shipping can no longer be determined by scale alone. The ability of the sector to leverage relevant technological advances to improve processes and operations, cut costs and generate value for the industry and customers, as well as the broader economy and society, is becoming increasingly important.

While the next chapters will address in more detail some of the implications of selected technologies, including for the world fleet, markets, ports and the regulatory framework, on the demand side and in connection with seaborne trade, the impact of digitalization can be significant, depending on the pace at which these technologies are implemented in shipping, the level of exposure of each market segment and the ability to strike a balance between the pros (for example, greater efficiency) and cons (for example, cybersecurity risks) associated with the various technologies. The challenge is to embrace the change while minimizing disruptions and supporting a sustainable recovery in shipping and global seaborne trade.

Based on these considerations, the following recommendations are suggested with a view to ensuring a more sustainable economic recovery in trade and shipping:

- Governments have a role to play by supporting the current positive economic trends and promoting a self-sustaining global economic recovery. This may entail, among other measures, actively promoting economic diversification in commodity-dependent countries. More importantly, at a time of growing concerns over the rise of protectionist sentiment, barriers to trade and trade disputes that may result in far-reaching detrimental impacts for the global economy and trade should be avoided to the extent possible.
- Relevant regulatory authorities, maritime transport analysts, as well as development entities such as UNCTAD need to regularly monitor market concentration trends in liner shipping and assess potential implications in terms of market power, freight rates, surcharges and other costs to shippers and trade.
- Governments, in collaboration with the shipping industry, the private sector, and the trade and business community need to build digital preparedness and promote greater uptake of relevant technologies. This will require, among others, providing an enabling legal and regulatory framework and supporting training and initiatives to build knowledge and upgrade skills.
- All stakeholders, including Governments, need to work together and support the development of transportation and supply chain infrastructure and services tailored for e-commerce. This may require an assessment of how the maritime transport sector could improve and tailor its service offerings to remain relevant and capture the potential gains deriving from e-commerce flows. A first step in this respect, is to enhance understanding of the cross-border e-commerce market and its potential. The establishment of a working group on measuring e-commerce and the digital economy, as proposed at the second session of the Intergovernmental Group of Experts on E-commerce and the Digital Economy, held in Geneva, Switzerland, in April 2018, could help promote a data-driven e-commerce analysis.
- While bearing in mind the need to prevent the heightened commoditization of services and ensure the ability to compete on service offerings to better respond to customer needs, collaboration between shipping lines, alliances, port terminals, shippers and other supply chain partners to improve communications, enhance transparency, increase efficiency, reduce operational complexity and allow better service offerings should be encouraged.

# REFERENCES

Asian Development Bank (2017). Changing patterns of trade and global value chains in post-crisis Asia. Asian Development Bank Briefs No. 76. February.

Barry Rogliano Salles (2018). Annual review 2018: Shipping and shipbuilding markets. Available at https://it4v7.interactiv-doc.fr/html/brsgroup2018annualreview_pdf_668.

Berenberg and Hamburg Institute of International Economics (2018). Strategy 2030: Shipping in an era of digital transformation. Available at www.berenberg.de.

British Petroleum (2018). *BP Statistical Review of World Energy 2018: June 2018* (Pureprint Group, London).

Brookings Institution (2018). Strengthening regional value chains: What's the role of the African Continental Free Trade Agreement? Africa in Focus. 21 March.

Clarksons Research (2018a). *Shipping Review and Outlook*. Spring.

Clarksons Research (2018b). *Seaborne Trade Monitor*. Volume 5. No. 5.

Clarksons Research (2018c). *China Intelligence Monthly*. April.

Clarksons Research (2018d). *Dry Bulk Trade Outlook*. Volume 24. No. 5. May.

Clarksons Research (2018e). *Container Intelligence Monthly*. Volume 20. No. 4. April.

Clarksons Research (2018f). *2018 'Trade Friction' Update*. June.

Colliers International (2017). Supply chain disruptors: Reshaping the supply chain. Quarter 2.

Danish Ship Finance (2017). *Shipping Market Review*. November.

Economic Commission for Latin America and the Caribbean (2010). Global Insight database.

Hellenic Shipping News (2017). China's Belt and Road Initiative: Rearranging global shipping? 6 June.

Horner R (2016) A new economic geography of trade and development? Governing South–South trade, value chains and production networks. *Territory, Politics, Governance*. 4(4):400-420.

International Monetary Fund (2016). Global trade: What's behind the slowdown? In: *World Economic Outlook: Subdued Demand – Symptoms and Remedies* (Washington, D.C.).

International Monetary Fund (2018). World Economic Outlook database. April.

Marine and Offshore Technology (2017). Digitalization in shipping is here to stay. 18 December.

McKinsey and Company (2017a). The alliance shuffle and consolidation: Implications for shippers. January.

McKinsey and Company (2017b). Container shipping: The next 50 years. October.

MDS Transmodal (2018). World Cargo Database. March.

Southern Africa Shipping News (2017). Container sector sees uptick in intra-Africa trade. 22 May.

Maersk (2018). Becoming the global integrator of container logistics. 9 February.

JOC.com (2017). Ocean freight to be a critical link in e-commerce supply chains. 17 May.

UNCTAD (2016). *Review of Maritime Transport 2016* (United Nations publication. Sales No. E.16.II.D.7, New York and Geneva).

UNCTAD (2017a). *Information Economy Report 2017: Digitalization, Trade and Development* (United Nations publication, Sales No. E.17.II.D.8, New York and Geneva).

UNCTAD (2017b). *Trade and Development Report 2017: Beyond Austerity – Towards a Global New Deal* (United Nations publication, Sales No. E.17.II.D.5, New York and Geneva).

UNCTAD (2018a). *Trade and Development Report 2018: Power, Platforms and the Free Trade Delusion* (United Nations publication, Sales No. E.18.II.D.7, New York and Geneva).

UNCTAD (2018b). UNCTADstat database. International trade.

UNCTAD (2018c). *World Investment Report 2018: Investment and New Industrial Policies* (United Nations publication, Sales No. E.18.II.D.4, New York and Geneva).

UNCTAD (2018d). Risks and benefits of data-driven economics in focus of major United Nations gathering. Press release. 28 March.

UNCTAD (2018e). African Continental Free Trade Area: Challenges and opportunities of tariff reductions. UNCTAD Research Paper No. 15.

United Nations (2018). *World Economic Situation and Prospects: Update as of Mid-2018*. New York.

World Steel Association (2018a). World crude steel output increases by 5.3% in 2017. 24 January.

World Steel Association (2018b). Global steel continues its broad recovery. 17 April.

World Trade Organization (2018). Strong trade growth in 2018 rests on policy choices. Press release 820. 12 April.

# ENDNOTES

1. Detailed figures on dry bulk commodities are derived from Clarksons Research, 2018d.
2. Free Trade Agreement between the European Union and Singapore; Investment Protection Agreement between the European Union and its Member States, of the One Part, and Singapore, of the Other Part.

After five years of decelerating growth, world fleet expansion increased slightly in 2017. A total of 42 million gross tons were added to the global tonnage in 2017, equivalent to a modest 3.3 per cent growth rate. This performance reflects both a slight upturn in new deliveries and a decrease in demolition activity, resulting from optimistic views among shipowners given positive developments in demand and freight rates. The expansion in ship supply capacity was surpassed by faster growth in demand and seaborne trade volumes, altering the market balance and supporting improved freight rates and earnings.

With regard to the shipping value chain, Germany remained the largest container ship owning country, although with a slight decrease in its share in 2017. By contrast, shipowners from Canada, China and Greece increased their container ship market shares. Further, the Marshall Islands emerged as the second largest registry, after Panama and ahead of Liberia. Over 90 per cent of shipbuilding activity in 2017 occurred in China, the Republic of Korea and Japan, and 79 per cent of ship demolitions took place in South Asia, notably in India, Bangladesh and Pakistan.

The liner shipping industry witnessed further consolidation through mergers and acquisitions and the restructuring of global alliances. However, despite the global trend in market concentration, UNCTAD data recorded an increase in 2017–2018 in the average number of companies providing services by country. This is the first such increase since UNCTAD began to monitor capacity deployment in 2004. Put differently, several individual carriers, both within and outside alliances, expanded their service networks to a larger number of countries, and this more than offset the reduction in the global number of companies following takeovers and mergers.

Not all countries saw an increase in the number of companies, however. UNCTAD data shows that the number of operators servicing several small island developing States and vulnerable economies decreased in 2017–2018. Further, reflecting the challenges posed by larger vessel sizes, small ports in many countries face obstacles in accommodating the demands of larger vessels and continue to rely on outdated and geared container and general cargo ships.

Three global liner shipping alliances dominate capacity deployment on the major container routes. The members of the alliances still compete with regard to prices, and the gains in operational efficiency and capacity utilization have exercised downward pressures on freight rates, to the benefit of shippers (see chapter 3). By joining forces in alliances, carriers have strengthened their bargaining power with regard to seaports when negotiating port calls and terminal operations (see chapter 4).

# WORLD FLEET

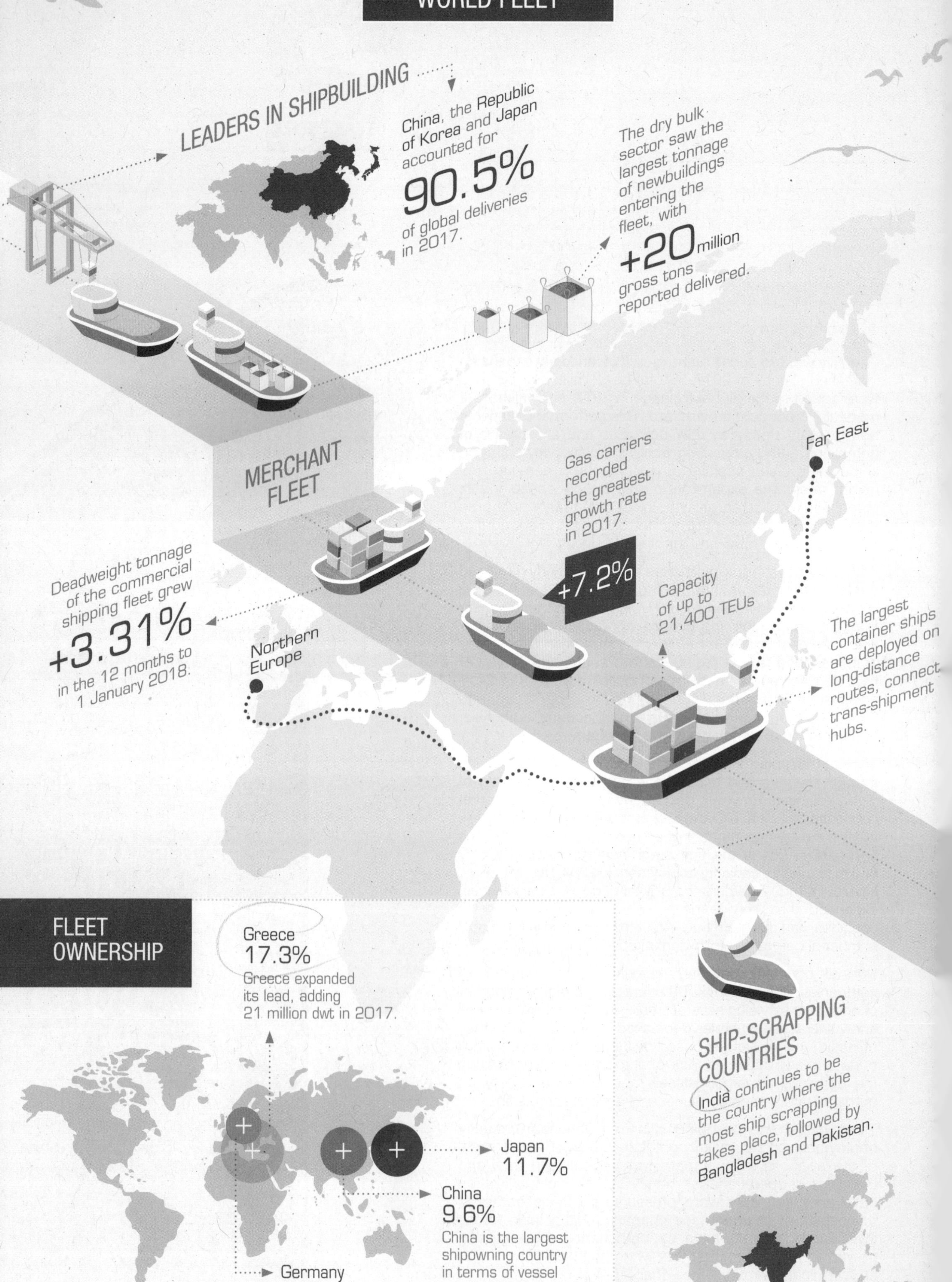

LEADERS IN SHIPBUILDING
China, the Republic of Korea and Japan accounted for
90.5%
of global deliveries in 2017.
The dry bulk sector saw the largest tonnage of newbuildings entering the fleet, with
+20 million
gross tons reported delivered.
MERCHANT FLEET
Gas carriers recorded the greatest growth rate in 2017.
+7.2%
Far East
Deadweight tonnage of the commercial shipping fleet grew
+3.31%
in the 12 months to 1 January 2018.
Capacity of up to 21,400 TEUs
Northern Europe
The largest container ships are deployed on long-distance routes, connect trans-shipment hubs.
FLEET OWNERSHIP
Greece
17.3%
Greece expanded its lead, adding 21 million dwt in 2017.
Japan
11.7%
China
9.6%
China is the largest shipowning country in terms of vessel numbers.
Germany
5.6%
SHIP-SCRAPPING COUNTRIES
India continues to be the country where the most ship scrapping takes place, followed by Bangladesh and Pakistan.

# A. WORLD FLEET STRUCTURE

Chapter 1 highlighted the demand side of and growth in seaborne trade volumes, which may serve as a leading indicator of or proxy for globalization, economic growth and merchandise trade expansion. However, such exchanges would not be possible without shipping and associated services, which provide in particular the global fleet of different vessels that cater for every type of cargo transported across the oceans. If seaborne trade volume is a proxy for the well-being of the global economy, the world fleet and the industry that provides the necessary vessels and services are the backbones of that economy. Beyond carrying 80 per cent of global trade by volume, ships also provide livelihoods for a wide range of businesses in nearly all countries of the world.

## 1. World fleet growth and principal vessel types

### *Growth in supply*

On 1 January 2018, the world commercial fleet consisted of 94,171 vessels, with a combined tonnage of 1.92 billion dwt. After five years of decelerating growth, in 2017, there was a slight rebound in the rate of increase (figure 2.1). The dead-weight tonnage of the commercial shipping fleet grew by 3.31 per cent in the 12 months to 1 January 2017, up from 3.15 per cent in 2016. Compared with the growth rate of demand, at 4.0 per cent in 2017, the lower level of growth in supply helped to improve market fundamentals, leading to improved freight rates and profits for most carriers, with the exception of tankers.

Ship sizes of new deliveries continued to be larger than the existing fleet. With regard to vessel numbers, the growth rate was therefore lower, at 1 per cent. The estimated market value of the world fleet, however, increased by 7.8 per cent, in line with improved market fundamentals and increased investments in ships incorporating the latest technologies and complying with current and potential future regulations.

### *Vessel types*

Dry bulk carriers, which carry iron ore, coal, grain and similar cargo, account for the largest share of the world fleet in dead-weight tonnage and the largest share of total cargo-carrying capacity, at 42.5 per cent (figure 2.2). They are followed by oil tankers, which carry crude oil and its products, and account for 29.2 per cent of total dead-weight tonnage. The third largest fleet is container ships, which account for 13.1 per cent of the total. As container ships carry goods of higher unit value than dry and liquid bulk ships and usually travel at higher speeds, they effectively carry more than half of total seaborne trade by monetary value.

**Figure 2.1** **Annual growth of world fleet and seaborne trade, 2000–2017**
(Percentage)

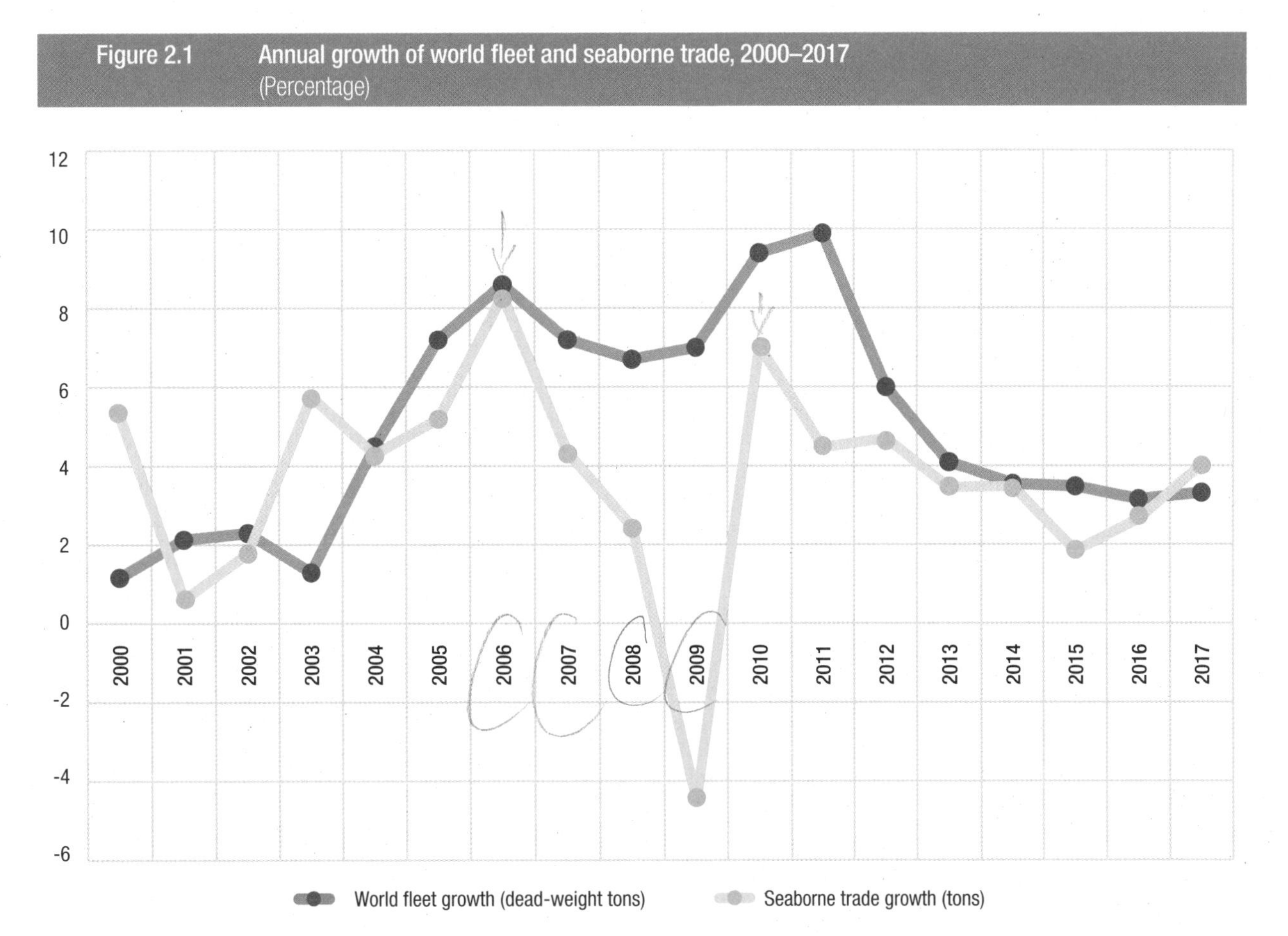


*Source:* UNCTAD, *Review of Maritime Transport*, various issues.

**Figure 2.2** Share of world fleet in dead-weight tonnage by principal vessel type, 1980–2018
(Percentage)

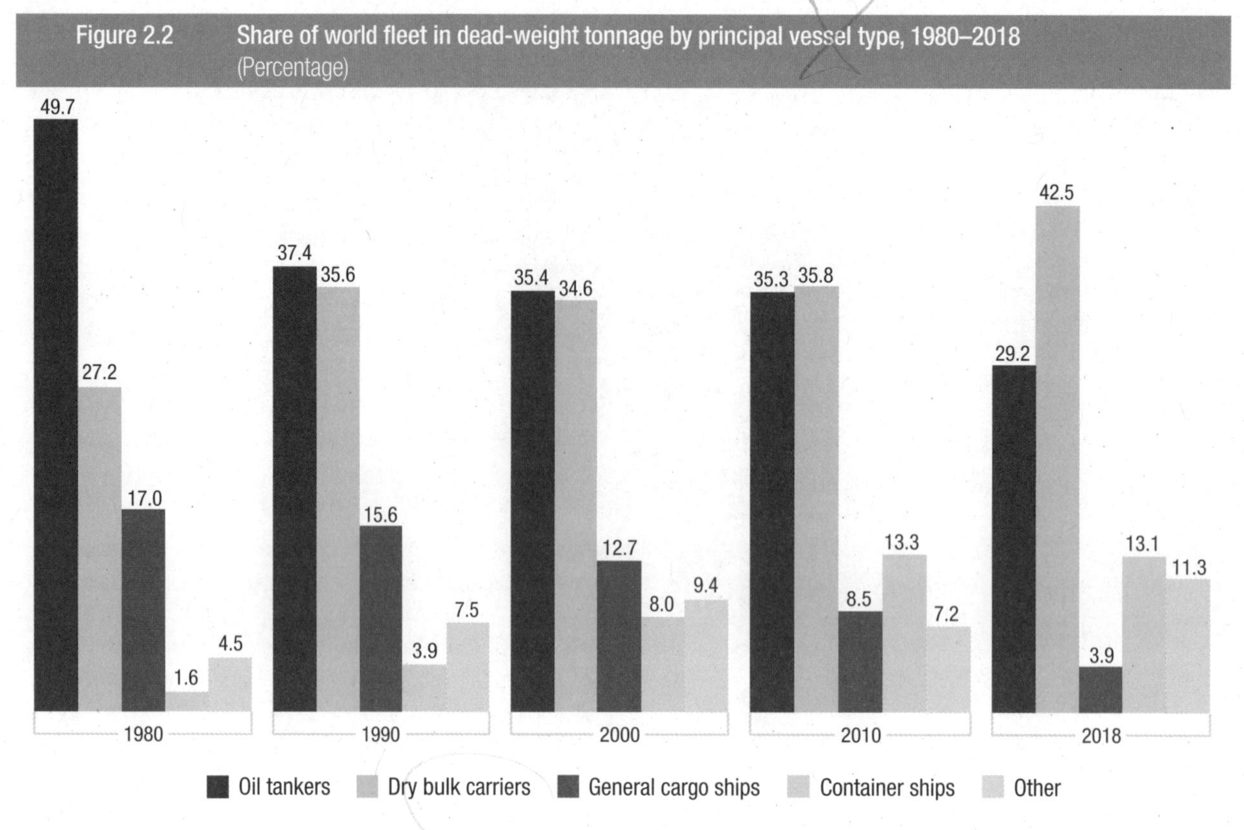


*Sources:* UNCTAD secretariat calculations, based on data from Clarksons Research and the *Review of Maritime Transport*, various issues.
*Notes:* Propelled seagoing merchant vessels of 100 gross tons and above, as at 1 January, excluding inland waterway vessels, fishing vessels, military vessels, yachts and offshore fixed and mobile platforms and barges, with the exception of floating production, storage and offloading units and drillships.

**Table 2.1** World fleet by principal vessel type, 2017–2018
(Thousands of dead-weight tons and percentage)

| | 2017 | 2018 | Percentage change, 2017–2018 |
|---|---|---|---|
| Oil tankers | 535 700 | 561 079 | 4.74 |
| | *28.8* | *29.2* | |
| Dry bulk carriers | 795 518 | 818 612 | 2.90 |
| | *42.7* | *42.5* | |
| General cargo ships | 74 908 | 74 458 | -0.60 |
| | *4.0* | *3.9* | |
| Container ships | 245 759 | 252 825 | 2.88 |
| | *13.2* | *13.1* | |
| Other | 210 455 | 217 028 | 3.12 |
| | *11.3* | *11.3* | |
| Gas carriers | 60 003 | 64 317 | 7.19 |
| | *3.2* | *3.3* | |
| Chemical tankers | 42 853 | 44 597 | 4.07 |
| | *2.3* | *2.3* | |
| Offshore vessels | 77 845 | 78 228 | 0.49 |
| | *4.2* | *4.1* | |
| Ferries and passenger ships | 5 944 | 6 075 | 2.20 |
| | *0.3* | *0.3* | |
| Other/not available | 23 810 | 23 811 | 0.01 |
| | *1.3* | *1.2* | |
| **World total** | **1 862 340** | **1 924 002** | **3.31** |

*Source:* UNCTAD secretariat calculations, based on data from Clarksons Research.
*Notes:* Propelled seagoing merchant vessels of 100 gross tons and above, as at 1 January. Percentage share in italics.

In 2017, almost all vessel types recorded positive growth rates, except for general cargo ships, which continued to show a long-term decline in their share of the world fleet (table 2.1). In January 2018, general cargo ships accounted for only 3.9 per cent of total dead-weight tonnage, a further decrease from their 4 per cent share in 2017. The long-term trend towards the containerization of general cargo may be illustrated by comparing the general cargo fleet with the container ship fleet. In 1980, container ships had one tenth the total tonnage of general cargo ships; at present, container ships have 3.4 times more total dead-weight tonnage. The order book for general cargo ships is at its lowest level since UNCTAD began to monitor this indicator and 58.8 per cent of such ships are older than 20 years (table 2.2).

Whenever there is sufficient volume, it is more efficient to make use of specialized ships for different types of cargo. General cargo ships therefore only remain in use in smaller markets, including at peripheral ports and on small islands and for shipments of project cargo that cannot be containerized. As the general cargo fleet continues to diminish, policymakers and port planners need to take every opportunity to invest in the most appropriate specialized terminals, in particular for the growing fleet of gearless container ships. A related development is the growing predominance of deep-water container trans-shipment hubs in all regions, which leads to a reduction in direct calls in adjacent smaller economies.

Gas carriers recorded the greatest growth rate in 2017, at 7.2 per cent, with expectations for further expansion in the coming years in view of the projected growth in liquefication and regasification capacity, as well as the consideration of gas as a cleaner source of energy. The share of chemical tankers grew by 4.1 per cent, reflecting the demand for the transport of chemicals required in industrial processing, as well as of palm oil and other liquid goods. The largest number of chemical tankers is controlled by owners from Japan, followed by owners from China, Norway, the Republic of Korea and Singapore.

### *Tonnage and value*

UNCTAD analysis mostly focuses on dead-weight tonnage, which is more relevant to seaborne trade and cargo-carrying capacity. To complement information on the maritime industry as a business sector, data on the commercial value of fleets are also included, indicating the capital intensiveness of the shipping industry and the implications for owning, operating, registering, building and scrapping such assets (figure 2.3). The value of its main assets also signals the state of the industry during business cycles. In addition, the value of a ship gives some indication of the level of its sophistication and technological content. For example, ships emit different amounts of greenhouse gases by ton-mile, depending on the country of build and vessel type (Right Ship, 2018). In the longer term, further digital transformation may entail greater investment and higher fixed costs, against lower operational and variable costs (box 2.1).

The high commercial value of the industry's main assets highlights the extent of investment in ships and technology, which shipowners need to recover by improving cost-efficiency measures, setting rates and surcharges and covering variable costs and fixed costs with regard to vessel prices. The values of different vessel types vary considerably (figure 2.3). Dry and liquid bulk ships have the largest cargo-carrying capacity and, accordingly, dry bulk carriers and oil tankers together account for more than 72 per cent of total dead-weight

**Box 2.1 The shipping fleet and digitalization**

The shipping industry is investing heavily in technologies that have the potential to transform business as usual. Such new technologies relate to the way that ships move and operate, as well as to strategic decision-making and day-to-day operations at offices, and include automated navigation and cargo-tracking systems and digital platforms that facilitate operations, trade and the exchange of data. They can potentially reduce costs, facilitate interactions between different actors and raise the maritime supply chain to the next level.

Automation and unstaffed ships offer interesting options related to greater cargo intake and reduced fuel consumption and operational expenses such as crew costs. At the same time, as new technologies are incorporated into on-board operations, ships become more complex to operate. As ship sizes and the complexity of on-board operations increase, the risk of major accidents may also rise. Yet reducing human intervention can also lead to a decrease in accidents. Human error reportedly accounted for approximately 75 per cent of the value of almost 15,000 marine liability insurance claims in 2011–2016, equivalent to over $1.6 billion.

Vessel and cargo-tracking systems are developing quickly. Technological developments can help in generating business intelligence for asset management and optimized operations, for example in the provision of data on fuel consumption and engine performance. Such systems also allow for the identification and monitoring of a ship's position, as well as for the monitoring of other aspects that might be important with regard to manoeuvring and stabilizing route and course, improving security and ensuring the safety of crew.

Combining on-board systems and digital platforms allows vessels and cargo to become a part of the Internet of things. A key challenge is to establish interoperability, so that data can be exchanged seamlessly, at the same time ensuring cybersecurity and the protection of commercially sensitive and private data (for further discussion of legal and regulatory frameworks, see chapter 5).

*Sources:* Allianz Global Corporate and Specialty, 2017; Lehmacher, 2017.

Figure 2.3 World fleet by principal vessel type, 2018
(Percentage)

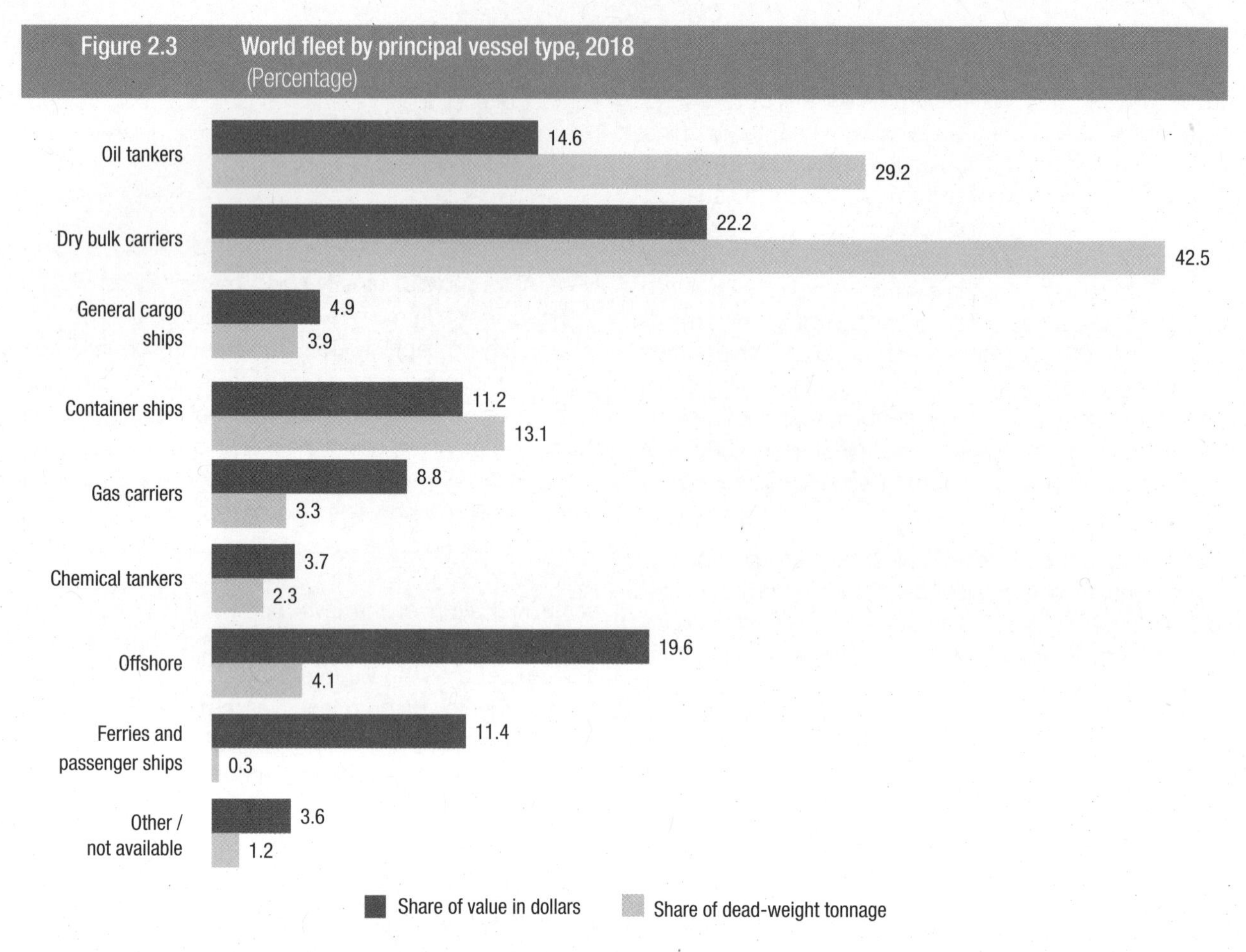


*Source:* UNCTAD secretariat calculations, based on data from Clarksons Research.
*Notes:* Share of dead-weight tonnage is calculated for all ships of 100 gross tons and above. Share of value is estimated for all commercial ships of 1,000 gross tons and above.

tonnage. However, with regard to their value, the vessels make up only 37 per cent of the fleet. Other vessel types are more technology-intensive and costlier to build. Gas carriers and the offshore fleet have a far higher monetary value by dwt. The category of ferries and passenger ships includes cruise ships and other vessels whose main purpose is not the transport of goods; their share in dead-weight tonnage is thus negligible, yet reaches more than 11 per cent of the fleet's market value.

## 2. World merchant fleet age distribution

The age structure of the world fleet provides interesting insights into trends and differences in country groups and vessel types with regard to fleet modernization and vessel sizes. The average age of the fleet registered in developing countries continues to be slightly higher than that registered in developed countries, but this gap has been narrowing over the years (table 2.2).

In 2017, as new deliveries further slowed down compared with deliveries in 2016, the average age of the world fleet increased slightly. At the beginning of 2018, the average vessel age in the commercial fleet was 20.8 years. With regard to dead-weight tonnage, the average age of the fleet was significantly younger, at 10.1 years, as ships built in the last 10 years have been on average seven times larger than those built two or more decades ago and still trading.

Container ship sizes have significantly increased in the last two decades, while the average size of oil tankers has marginally decreased. The largest ships built in the last five years have been container ships of an average of 83,122 dwt, followed by dry bulk carriers of an average of 79,281 dwt. These trends are a reflection of changed economic conditions. Notably, in container shipping, the process of consolidation has gone together with the demand for larger ships by the major shipping lines and alliances.

## 3. Container ship fleet

Container shipping is fundamental for global trade in intermediate and manufactured consumer goods. It is provided by regular liner shipping services that form a network of transport connections, including direct services and services that involve the trans-shipment of containers in hub ports.

Modern container ports have specialized ship-to-shore container cranes installed and most new container ships are therefore gearless, that is, they are not equipped with

| Table 2.2 | Age distribution of world merchant fleet by vessel type, 2018 | | | | | | | | |
|---|---|---|---|---|---|---|---|---|---|
| **Economic grouping and vessel type** | | **Years** | | | | | **Average age** | | **Percentage change** |
| | | **0–4** | **5–9** | **10–14** | **15–19** | **20+** | **2018** | **2017** | **2017–2018** |
| **World** | | | | | | | | | |
| Oil tankers | Percentage of total ships | 14.97 | 21.89 | 17.04 | 8.46 | 37.64 | 19.06 | 18.73 | 0.32 |
| | Percentage of dead-weight tonnage | 21.70 | 33.86 | 24.60 | 14.30 | 5.55 | 9.99 | 9.90 | 0.09 |
| | Average vessel size (dwt) | 78 543 | 84 016 | 78 643 | 93 525 | 8 303 | | | |
| Dry bulk carriers | Percentage of total ships | 27.83 | 41.32 | 12.90 | 8.72 | 9.24 | 9.10 | 8.77 | 0.33 |
| | Percentage of dead-weight tonnage | 29.99 | 43.04 | 12.93 | 7.22 | 6.82 | 8.28 | 7.93 | 0.34 |
| | Average vessel size (dwt) | 79 281 | 76 618 | 73 750 | 60 907 | 54 304 | | | |
| General cargo ships | Percentage of total ships | 6.09 | 16.26 | 11.88 | 7.03 | 58.75 | 25.82 | 25.10 | 0.72 |
| | Percentage of dead-weight tonnage | 11.59 | 26.27 | 14.50 | 9.84 | 37.80 | 18.66 | 18.17 | 0.49 |
| | Average vessel size (dwt) | 8 060 | 6 641 | 5 400 | 6 392 | 2 656 | | | |
| Container ships | Percentage of total ships | 17.40 | 26.67 | 26.81 | 14.74 | 14.37 | 11.94 | 11.53 | 0.41 |
| | Percentage of dead-weight tonnage | 29.55 | 30.98 | 23.71 | 10.32 | 5.45 | 9.04 | 8.71 | 0.32 |
| | Average vessel size (dwt) | 83 122 | 56 847 | 43 284 | 34 246 | 18 568 | | | |
| Other | Percentage of total ships | 13.07 | 19.42 | 11.62 | 8.48 | 47.41 | 22.86 | 22.32 | 0.54 |
| | Percentage of dead-weight tonnage | 20.70 | 24.04 | 16.10 | 10.78 | 28.39 | 15.45 | 15.34 | 0.11 |
| | Average vessel size (dwt) | 9 253 | 7 507 | 8 440 | 7 741 | 4 156 | | | |
| All ships | Percentage of total ships | 13.75 | 22.01 | 13.25 | 8.54 | 42.46 | 20.83 | 20.34 | 0.50 |
| | Percentage of dead-weight tonnage | 25.74 | 35.98 | 18.16 | 10.20 | 9.92 | 10.09 | 9.85 | 0.24 |
| | Average vessel size (dwt) | 43 360 | 38 186 | 32 634 | 29 049 | 6 150 | | | |
| **Developing economies – all ships** | | | | | | | | | |
| | Percentage of total ships | 14.08 | 22.81 | 12.70 | 7.76 | 42.65 | 20.07 | 19.56 | 0.51 |
| | Percentage of dead-weight tonnage | 25.70 | 35.39 | 13.92 | 10.03 | 14.97 | 17.46 | 17.50 | -0.04 |
| | Average vessel size (dwt) | 34 174 | 30 399 | 21 763 | 25 426 | 6 932 | | | |
| **Developed economies – all ships** | | | | | | | | | |
| | Percentage of total ships | 14.58 | 23.78 | 15.57 | 10.63 | 35.45 | 19.35 | 18.94 | 0.41 |
| | Percentage of dead-weight tonnage | 26.15 | 36.71 | 20.97 | 10.26 | 5.91 | 9.35 | 9.12 | 0.23 |
| | Average vessel size (dwt) | 55 976 | 47 322 | 43 041 | 32 571 | 6 951 | | | |
| **Transition economies – all ships** | | | | | | | | | |
| | Percentage of total ships | 5.75 | 9.48 | 6.81 | 3.54 | 74.41 | 29.67 | 29.08 | 0.59 |
| | Percentage of dead-weight tonnage | 9.80 | 27.51 | 22.07 | 13.44 | 27.18 | 16.16 | 15.55 | 0.62 |
| | Average vessel size (dwt) | 13 865 | 22 668 | 25 258 | 26 867 | 2 577 | | | |

*Source:* UNCTAD secretariat calculations, based on data from Clarksons Research.
*Notes:* Propelled seagoing vessels of 100 gross tons and above, as at 1 January.

**Figure 2.4 Container ship deliveries, 2005–2017**
(20-foot equivalent units)

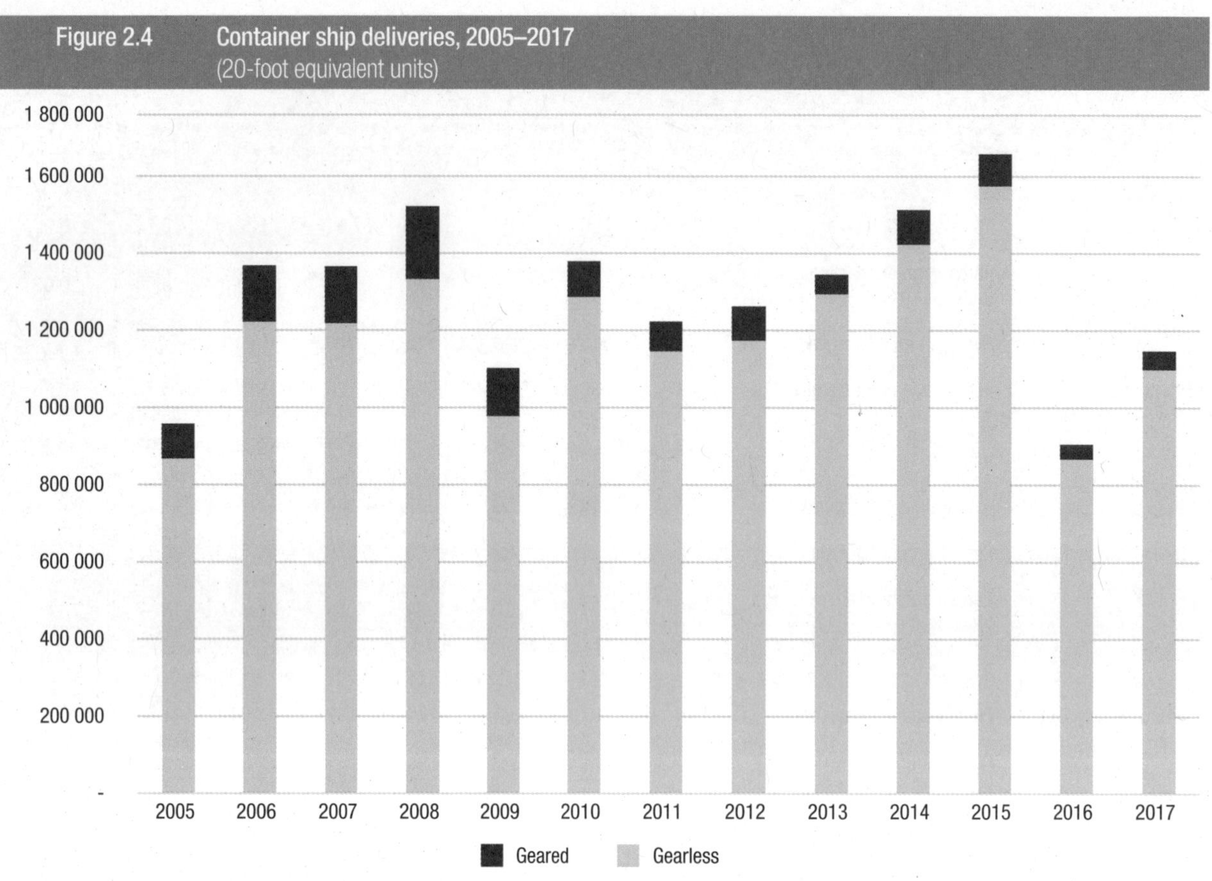


*Source:* UNCTAD secretariat calculations, based on data from Clarksons Research.
*Note:* Propelled seagoing vessels of 100 gross tons and above.

**Figure 2.5 Trends in container ship deployment, average per country**
(2004 = 100)

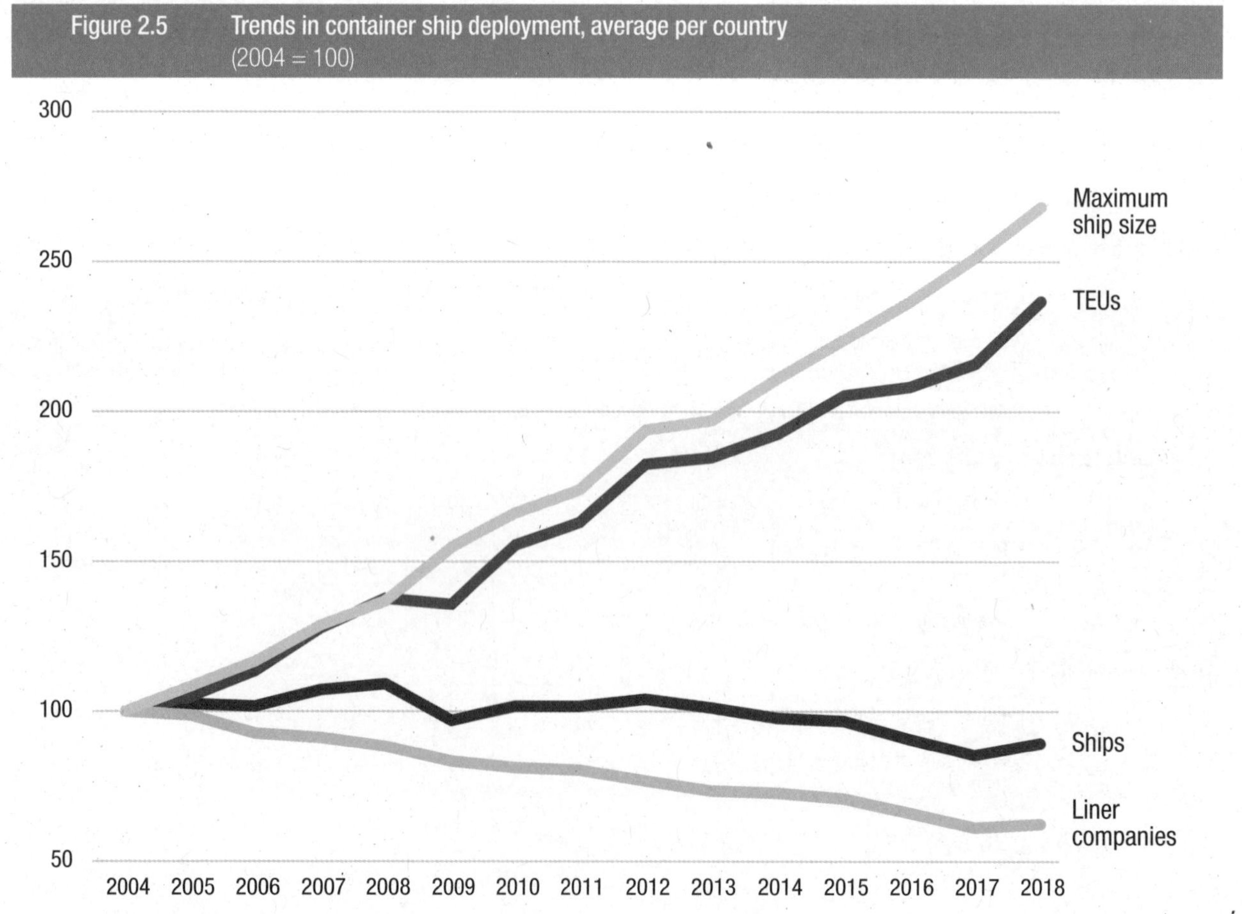


*Source:* UNCTAD secretariat calculations, based on data from MDS Transmodal and *Review of Maritime Transport*, various issues.
*Note:* Figures as at 1 May of each year.

their own cranes. In 2017, only 4.2 per cent of TEUs of container ship deliveries was of geared container ships, intended for markets in which terminals do not provide for the necessary port cranes, including in some small island developing States and at small and remote ports at which the volume of cargo may not justify investment in ship-to-shore cranes (figure 2.4).

With regard to long-term trends in container ship deployment by country, ship sizes and total capacity deployed by country have increased over the years and the number of companies has decreased (figure 2.5). The number of ships and TEU-carrying capacity deployed reflect to some extent the growth of containerized trade. For example, deployment declined in 2008–2009, following the economic crisis, when carriers withdrew capacity from the market. The latest developments are more positive and the average TEU deployment by country increased by almost 10 per cent between May 2017 and May 2018. However, the number of companies providing services to and from a country, on average, has decreased in most years since 2004. The slight increase between 2017 and 2018 is an interesting development, as it reflects the fact that despite global mergers and acquisitions, the remaining carriers have been expanding into new markets, including as members of global alliances. Each major carrier thereby ensures its own in-house global network.

The largest ships are deployed on the Far East–Northern Europe route. As at June 2018, there were 18 weekly services on this route, down from 32 services in 2008, when significantly smaller ships were deployed. Current services are operated by nine different carriers organized into three alliances and one independent carrier, Hyundai Merchant Marine, and the average capacity of the total 205 ships employed is 15,000 TEUs; the largest vessel has a capacity of 21,400 TEUs and the smallest vessel, deployed by the sole independent carrier, has a capacity of 4,100 TEUs (Dynamar BV, 2018a).

The slight long-term decline in the number of ships deployed by country does not mean that the total number of ships in the world fleet has declined. The opposite is true; the total number of container ships in the world fleet increased in 2004–2018. Each ship calls at a smaller number of ports; the largest ships are deployed on long-distance routes, connecting trans-shipment hubs, and the smaller ships connect a smaller number of countries, on shorter routes, to and from these trans-shipment hubs.

## B. WORLD FLEET OWNERSHIP AND OPERATION

### 1. Shipowning countries

The top five shipowning countries together account for 49.6 per cent of the world fleet in dead-weight tonnage. Greece has expanded its lead, adding 21 million dwt in 2017; it now has a market share of 17.3 per cent, followed by Japan at 11.7 per cent, China at 9.6 per cent and Germany at 5.6 per cent. Shipowners from Greece specialize in oil tankers, in which Greece has a market share of 24 per cent, as well as dry bulk carriers. Japan and China have their largest market shares in dry bulk carriers, with 20 and 16 per cent, respectively. Shipowners from Germany specialize mostly in container ships, in which Germany has a market share of 20 per cent. Among charter owners, that is, owners that do not themselves provider liner services but instead charter ships to liner companies, Germany has a market share of one third, down from two thirds in 2013, and owners from Canada, China and Greece have expanded their markets. A typical example of this trend is the sale of six container ships by Commerzbank of Germany to Maersk in March 2018, for around $280 million (Dynamar BV, 2018b).

The largest shipowning country in terms of vessel numbers is China, with 5,512 commercial ships of 1,000 gross tons and above, many of which are deployed in domestic trades, under the national flag (table 2.3). Indonesia and the Russian Federation also own a large number of ships deployed in coastal and inter-island transport. Most major shipowning economies are in Asia, Europe and North America. No country in Africa or Oceania and only one country in Latin America – Brazil – is among the top 35 shipowners. Among the top 35 shipowning countries, 28 have more than half of their fleet registered abroad, that is, in a foreign open registry. The seven exceptions are Belgium, India, Indonesia, Italy, Saudi Arabia, Thailand and Viet Nam. In Saudi Arabia and Thailand, the nationally flagged ships are mostly oil tankers; in Belgium and Italy, the national flag is financially attractive for national owners; and in India, Indonesia and Viet Nam, the nationally flagged ships include a large share of general cargo ships deployed in coastal traffic, which is reserved for nationally flagged ships.

With regard to the commercial value of the world fleet, the largest shipowning country is the United States, followed by Japan and Greece (figure 2.6). The difference between the ranking by tonnage and by value is due to the vessel types owned by different countries. For example, shipowners from Greece specialize in dry bulk carriers and oil tankers, which have a large carrying capacity; shipowners from the United States, by contrast, have greater shares in cruise ships and other vessels, primarily offshore, which are not used for trade in goods.

### 2. Container ship ownership and operation

Table 2.4 depicts container ship fleet ownership in TEUs. Germany continues to be the largest owner, with a market share of 20.22 per cent, a decrease of 1.2 percentage points from 2017. France, Denmark, Hong Kong (China) and Switzerland own the container

**Table 2.3 Ownership of world fleet ranked by dead-weight tonnage, 2018**

| | Country or territory | Number of vessels | | | Dead-weight tonnage (thousands of tons) | | | |
|---|---|---|---|---|---|---|---|---|
| | | National flag | Foreign or international flag | Total | National flag | Foreign or international flag | Total | National flag as percentage of total (dead-weight tonnage) |
| 1 | Greece | 774 | 3 597 | 4 371 | 64 977 | 265 199 | 330 176 | 19.7 |
| 2 | Japan | 988 | 2 853 | 3 841 | 38 053 | 185 562 | 223 615 | 17.0 |
| 3 | China | 3 556 | 1 956 | 5 512 | 83 639 | 99 455 | 183 094 | 45.7 |
| 4 | Germany | 319 | 2 550 | 2 869 | 11 730 | 95 389 | 107 119 | 11.0 |
| 5 | Singapore | 240 | 2 389 | 2 629 | 2 255 | 101 327 | 103 583 | 2.2 |
| 6 | Hong Kong (China) | 95 | 1 497 | 1 592 | 2 411 | 95 396 | 97 806 | 2.5 |
| 7 | Republic of Korea | 801 | 825 | 1 626 | 14 019 | 63 258 | 77 277 | 18.1 |
| 8 | United States | 943 | 1 128 | 2 071 | 13 319 | 55 611 | 68 930 | 19.3 |
| 9 | Norway | 549 | 1 433 | 1 982 | 4 944 | 54 437 | 59 380 | 8.3 |
| 10 | Bermuda | 21 | 473 | 494 | 1 215 | 53 036 | 54 252 | 2.2 |
| 11 | Taiwan Province of China | 164 | 823 | 987 | 6 732 | 43 690 | 50 422 | 13.4 |
| 12 | United Kingdom | 398 | 956 | 1 354 | 9 496 | 40 494 | 49 989 | 19.0 |
| 13 | Monaco | 16 | 405 | 421 | 3 856 | 35 467 | 39 323 | 9.8 |
| 14 | Denmark | 139 | 805 | 944 | 1 521 | 37 691 | 39 212 | 3.9 |
| 15 | Turkey | 633 | 889 | 1 522 | 8 034 | 19 207 | 27 241 | 29.5 |
| 16 | India | 885 | 126 | 1 011 | 17 974 | 6 878 | 24 852 | 72.3 |
| 17 | Switzerland | 43 | 368 | 411 | 1 565 | 23 240 | 24 805 | 6.3 |
| 18 | Belgium | 120 | 152 | 272 | 12 405 | 11 225 | 23 630 | 52.5 |
| 19 | Russian Federation | 1 384 | 323 | 1 707 | 7 589 | 14 630 | 22 219 | 34.2 |
| 20 | Indonesia | 1 886 | 62 | 1 948 | 19 414 | 885 | 20 299 | 95.6 |
| 21 | Italy | 583 | 163 | 746 | 14 221 | 5 530 | 19 750 | 72.0 |
| 22 | Malaysia | 500 | 162 | 662 | 9 731 | 9 793 | 19 524 | 49.8 |
| 23 | Netherlands | 800 | 428 | 1 228 | 6 911 | 11 205 | 18 116 | 38.2 |
| 24 | Islamic Republic of Iran | 164 | 62 | 226 | 3 914 | 13 927 | 17 841 | 21.9 |
| 25 | United Arab Emirates | 200 | 695 | 895 | 1 115 | 16 317 | 17 432 | 6.4 |
| 26 | Saudi Arabia | 219 | 67 | 286 | 13 378 | 3 760 | 17 138 | 78.1 |
| 27 | France | 159 | 279 | 438 | 5 635 | 6 506 | 12 141 | 46.4 |
| 28 | Brazil | 290 | 100 | 390 | 4 341 | 7 636 | 11 976 | 36.2 |
| 29 | Cyprus | 14 | 281 | 295 | 92 | 10 137 | 10 229 | 0.9 |
| 30 | Viet Nam | 875 | 116 | 991 | 7 464 | 1 756 | 9 221 | 81.0 |
| 31 | Canada | 220 | 149 | 369 | 2 695 | 6 387 | 9 082 | 29.7 |
| 32 | Oman | 6 | 42 | 48 | 6 | 7 782 | 7 788 | 0.1 |
| 33 | Thailand | 337 | 65 | 402 | 5 576 | 1 983 | 7 559 | 73.8 |
| 34 | Qatar | 63 | 56 | 119 | 1 841 | 4 977 | 6 818 | 27.0 |
| 35 | Sweden | 167 | 122 | 289 | 2 332 | 3 927 | 6 259 | 37.3 |
| | **Subtotal, top 35 shipowners** | **18 551** | **26 397** | **44 948** | **404 399** | **1 413 699** | **1 818 098** | **22.2** |
| | *Rest of world and unknown* | *3 224* | *2 560* | *5 784* | *36 114* | *55 800* | *91 913* | *39.3* |
| | **World total** | **21 775** | **28 957** | **50 732** | **440 513** | **1 469 499** | **1 910 012** | **23.1** |

*Source:* UNCTAD secretariat calculations, based on data from Clarksons Research.

*Notes:* Propelled seagoing vessels of 1,000 gross tons and above, as at 1 January.

For a complete listing of nationally owned fleets, see http://stats.unctad.org/fleetownership.

For the purposes of this table, second and international registries are recorded as foreign or international registries, whereby, for example, ships of owners in the United Kingdom registered in Gibraltar or the Isle of Man are recorded as under a foreign or international flag. In addition, ships of owners in Denmark registered in the Danish International Register of Shipping account for 43.5 per cent of the Denmark-owned fleet in dead-weight tonnage and ships of owners in Norway registered in the Norwegian International Ship Register account for 26.4 per cent of the Norway-owned fleet in dead-weight tonnage.

*Abbreviation:* SAR, Special Administrative Region.

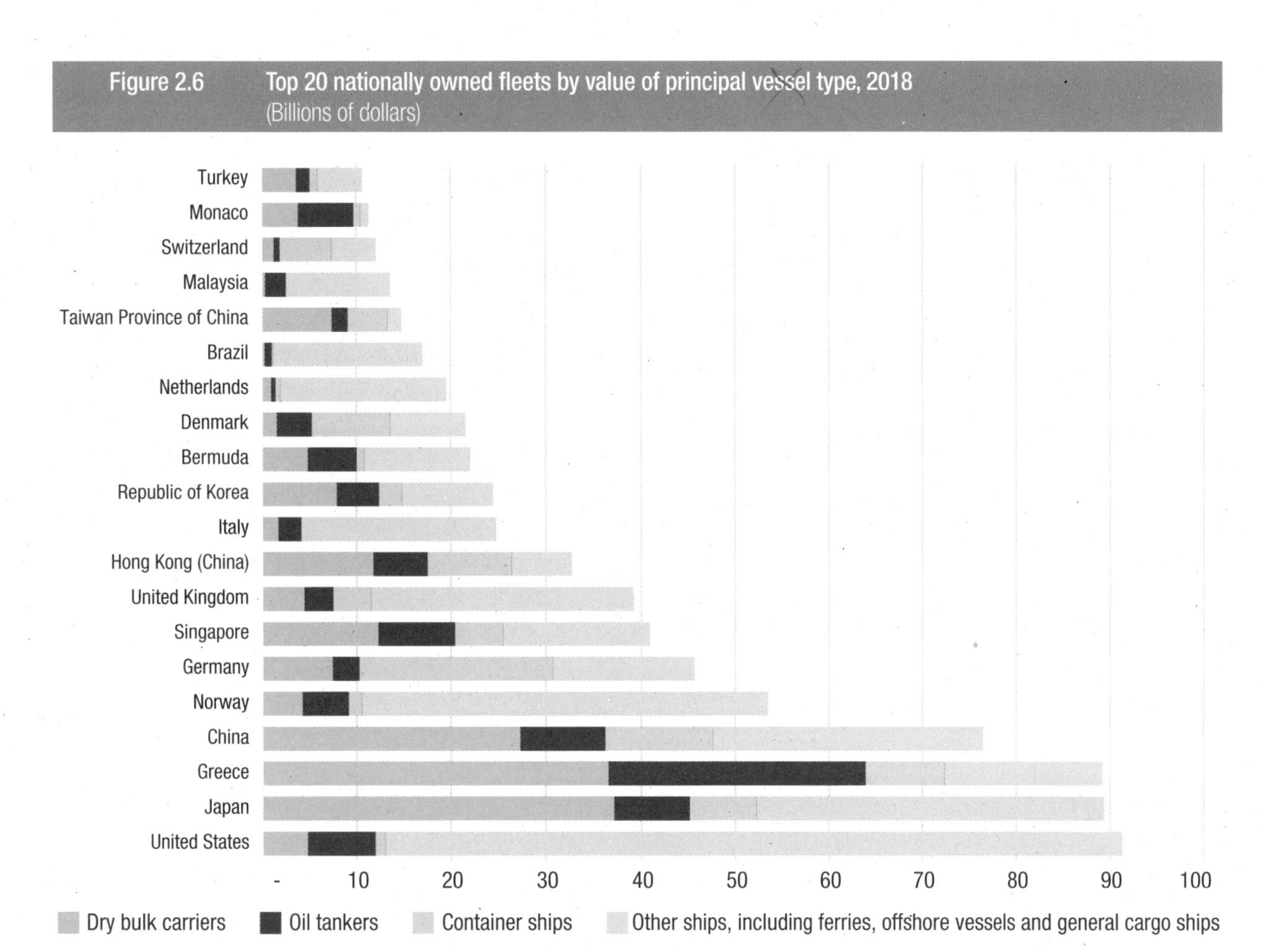


**Figure 2.6** Top 20 nationally owned fleets by value of principal vessel type, 2018
(Billions of dollars)

*Source:* UNCTAD secretariat calculations, based on data from Clarksons Research.
*Notes:* Propelled seagoing merchant vessels of 1,000 gross tons and above, as at 1 January.

**Table 2.4** Global top 20 owners of container-carrying world fleet, 2018

| Country or territory | 20-foot equivalent units | Market share (percentage) | Number of ships | Average age per ship (years) | Size of largest ship (20-foot equivalent units) | Average size per ship (20-foot equivalent units) |
|---|---|---|---|---|---|---|
| Germany | 4 207 388 | 20.22 | 1 131 | 10.6 | 18 800 | 3 720 |
| Denmark | 2 220 911 | 10.68 | 317 | 10.5 | 20 568 | 7 006 |
| China | 2 150 700 | 10.34 | 485 | 10.8 | 19 224 | 4 434 |
| Greece | 1 891 234 | 9.09 | 418 | 11.7 | 14 424 | 4 524 |
| Hong Kong (China) | 1 583 036 | 7.61 | 258 | 8.8 | 21 413 | 6 136 |
| Japan | 1 455 580 | 7.00 | 278 | 8.7 | 20 150 | 5 236 |
| Switzerland | 1 260 807 | 6.06 | 207 | 15.5 | 14 000 | 6 091 |
| France | 1 038 824 | 4.99 | 135 | 9.4 | 17 722 | 7 695 |
| Taiwan Province of China | 985 495 | 4.74 | 255 | 13.1 | 8 626 | 3 865 |
| United Kingdom | 870 632 | 4.18 | 199 | 10.8 | 15 908 | 4 375 |
| Singapore | 658 654 | 3.17 | 230 | 11.9 | 15 908 | 2 864 |
| Republic of Korea | 532 670 | 2.56 | 186 | 12.5 | 13 100 | 2 864 |
| Cyprus | 253 392 | 1.22 | 70 | 10.2 | 19 200 | 3 620 |
| Norway | 208 262 | 1.00 | 48 | 9.9 | 13 102 | 4 339 |
| United States | 207 894 | 1.00 | 70 | 19.4 | 9 443 | 2 970 |
| Indonesia | 172 711 | 0.83 | 205 | 17.4 | 3 534 | 842 |
| Israel | 170 434 | 0.82 | 31 | 8.7 | 10 062 | 5 498 |
| Turkey | 159 855 | 0.77 | 90 | 14.0 | 9 010 | 1 776 |
| United Arab Emirates | 110 265 | 0.53 | 61 | 17.0 | 4 498 | 1 808 |
| Netherlands | 92 815 | 0.45 | 87 | 10.8 | 3 508 | 1 067 |
| Subtotal, top 20 owners | 20 231 559 | 97.25 | 4 761 | 11.1 | 21 413 | 4 249 |
| *Rest of world* | *572 912* | *2.75* | *383* | *12.6* | *6 572* | *1 496* |
| **World total** | **20 804 471** | **100.00** | **5 144** | **11.9** | **21 413** | **2 004** |

*Source:* UNCTAD secretariat calculations, based on data from Clarksons Research.
*Notes:* Propelled seagoing vessels of 1,000 gross tons and above, as at 1 January. Only fully cellular container ships are included. For a complete listing of nationally owned fleets, see http://stats.unctad.org/fleetownership.
*Abbreviation:* SAR, Special Administrative Region.

**Table 2.5 Global top 30 liner shipping companies, 1 June 2018**

| | Owned | | | Chartered | | | Total | | | | | |
|---|---|---|---|---|---|---|---|---|---|---|---|---|
| | Number of ships | Total 20-foot equivalent units | Average vessel size (20-foot equivalent units) | Number of ships | Total 20-foot equivalent units | Average vessel size (20-foot equivalent units) | Number of ships | Total 20-foot equivalent units | Market share (percentage of 20-foot equivalent units) | Average vessel size (20-foot equivalent units) | Share of chartered ships (percentage) |
| Maersk | 300 | 2 213 253 | 7 378 | 400 | 1 666 186 | 4 165 | 700 | 3 879 439 | 15.3 | 5 542 | 42.9 |
| Mediterranean Shipping Company | 154 | 1 032 256 | 6 703 | 319 | 2 085 852 | 6 539 | 473 | 3 118 108 | 12.3 | 6 592 | 66.9 |
| CMA CGM | 147 | 1 131 606 | 7 698 | 329 | 1 422 658 | 4 324 | 476 | 2 554 264 | 10.1 | 5 366 | 55.7 |
| China Ocean Shipping (Group) Company | 156 | 1 194 776 | 7 659 | 174 | 777 715 | 4 470 | 330 | 1 972 491 | 7.8 | 5 977 | 39.4 |
| Hapag-Lloyd | 105 | 999 787 | 9 522 | 112 | 551 087 | 4 920 | 217 | 1 550 874 | 6.1 | 7 147 | 35.5 |
| Ocean Network Express | 88 | 700 560 | 7 961 | 140 | 835 752 | 5 970 | 228 | 1 536 312 | 6.1 | 6 738 | 54.4 |
| Evergreen | 113 | 577 062 | 5 107 | 87 | 533 646 | 6 134 | 200 | 1 110 708 | 4.4 | 5 554 | 48.0 |
| Orient Overseas Container Line | 55 | 495 150 | 9 003 | 44 | 194 836 | 4 428 | 99 | 689 986 | 2.7 | 6 970 | 28.2 |
| Yang Ming | 45 | 209 810 | 4 662 | 55 | 399 939 | 7 272 | 100 | 609 749 | 2.4 | 6 097 | 65.6 |
| Pacific International Lines | 118 | 348 140 | 2 950 | 14 | 65 194 | 4 657 | 132 | 413 334 | 1.6 | 3 131 | 15.8 |
| Zim Integrated Shipping Services | 11 | 70 314 | 6 392 | 72 | 328 612 | 4 564 | 83 | 398 926 | 1.6 | 4 806 | 82.4 |
| Hyundai Merchant Marine | 20 | 158 886 | 7 944 | 45 | 223 258 | 4 961 | 65 | 382 144 | 1.5 | 5 879 | 58.4 |
| Wan Hai Lines | 72 | 172 819 | 2 400 | 28 | 82 263 | 2 938 | 100 | 255 082 | 1.0 | 2 551 | 32.2 |
| X-Press Feeders | 20 | 17 253 | 863 | 69 | 109 462 | 1 586 | 89 | 126 715 | 0.5 | 1 424 | 86.4 |
| Republic of Korea Marine Transport Company | 27 | 57 082 | 2 114 | 30 | 67 378 | 2 246 | 57 | 124 460 | 0.5 | 2 184 | 54.1 |
| Islamic Republic of Iran Shipping Lines | 24 | 79 668 | 3 320 | 4 | 22 850 | 5 713 | 28 | 102 518 | 0.4 | 3 661 | 22.3 |
| Shandong International Transportation Corporation | 50 | 70 719 | 1 414 | 17 | 23 950 | 1 409 | 67 | 94 669 | 0.4 | 1 413 | 25.3 |
| SM Line | 13 | 57 706 | 4 439 | 7 | 20 612 | 2 945 | 20 | 78 318 | 0.3 | 3 916 | 26.3 |
| Arkas Line | 37 | 65 336 | 1 766 | 7 | 9 940 | 1 420 | 44 | 75 276 | 0.3 | 1 711 | 13.2 |
| TS Lines | 4 | 7 200 | 1 800 | 29 | 66 312 | 2 287 | 33 | 73 512 | 0.3 | 2 228 | 90.2 |
| Transworld Group of Companies | 22 | 38 159 | 1 735 | 11 | 22 302 | 2 027 | 33 | 60 461 | 0.2 | 1 832 | 36.9 |
| Feedertech Shipping | 5 | 12 040 | 2 408 | 12 | 44 422 | 3 702 | 17 | 56 462 | 0.2 | 3 321 | 78.7 |
| Grimaldi Group | 41 | 48 110 | 1 173 | 7 | 3 343 | 478 | 48 | 51 453 | 0.2 | 1 072 | 6.5 |
| Quanzhou Ansheng Shipping Company | 20 | 50 820 | 2 541 | | | | 20 | 50 820 | 0.2 | 2 541 | 0.0 |
| Regional Container Lines | 20 | 28 928 | 1 446 | 7 | 17 060 | 2 437 | 27 | 45 988 | 0.2 | 1 703 | 37.1 |
| Unifeeder | 1 | 530 | 530 | 38 | 42 883 | 1 129 | 39 | 43 413 | 0.2 | 1 113 | 98.8 |
| China Navigation Company | 19 | 31 872 | 1 677 | 6 | 10 859 | 1 810 | 25 | 42 731 | 0.2 | 1 709 | 25.4 |
| Grieg Star | 26 | 41 540 | 1 598 | 1 | 306 | 306 | 27 | 41 846 | 0.2 | 1 550 | 0.7 |
| Sinotrans | 13 | 21 102 | 1 623 | 13 | 20 139 | 1 549 | 26 | 41 241 | 0.2 | 1 586 | 48.8 |
| Sinokor Merchant Marine | 12 | 17 874 | 1 490 | 18 | 22 409 | 1 245 | 30 | 40 283 | 0.2 | 1 343 | 55.6 |
| Subtotal, top 30 carriers | 1 738 | 9 950 358 | 5 725 | 2 095 | 9 671 225 | 4 616 | 3 833 | 19 621 583 | 77.6 | 5 119 | 49.3 |
| Rest of world | | | | | | | 4 330 | 5 668 430 | 22.4 | 1 309 | |
| World total | | | | | | | 8 163 | 25 290 013 | 100.0 | 3 098 | |

*Source:* *UNCTAD* secretariat calculations, based on data from MDS Transmodal.

ships with the largest average size and also host the largest liner shipping companies, which tend to own the largest vessels. Smaller vessels are more likely to be chartered from owners in, for example, Germany and Greece. The top three carriers are from Europe, with a combined market share of 37.7 per cent of world carrying capacity. Most of the remaining top 30 carriers are from Asia. In total, the top 10 carriers have a combined market share of 68.6 per cent and the top 30 together account for 77.6 per cent (table 2.5). Carriers with more ships also own and operate larger ships, which is a further indication that the growing size of container ships and the process of consolidation go hand in hand.

The liner shipping industry has witnessed increasing consolidation, in the form of both mergers and acquisitions, and liner shipping alliances. Consolidation can result in better supply management, fleet utilization and improved efficiency. It can benefit the industry through the pooling of cargo, improved economies of scale and reduced operating costs. Carriers may also see the benefits of such cooperation by sharing resources, including port calls and networks, and developing new services. Shippers could benefit from consolidation through stability and less fluctuation in freight rates, as well as more efficient and extensive services offered by carriers. As long as there is sufficient competition and transparency, shippers may also benefit from improvements if the resulting lower costs are effectively passed on to them in the form of lower freight rates. Beyond cost savings, improvements in operational efficiency and higher vessel utilization can exacerbate the oversupply of capacity, leading to further downward pressure on freight rates.

Consolidation can have a potential negative impact on competition, however, and may result in oligopolistic market structures. Growing consolidation can reinforce market power, potentially leading to decreased supply and service quality and higher prices. Some of these negative outcomes may already be in effect. For example, in 2017–2018, the number of operators decreased in several small island developing States and structurally weak developing countries (table 2.6). This is an issue of concern, as such countries are already serviced by a low number of operators and face high transport costs due to several obstacles, including limited transport infrastructure and market size. Alliances have also increased the bargaining power of shipping companies with regard to ports. By pooling services and ship calls, for example when negotiating port dues or conditions for dedicated terminals, carriers can more easily obtain the most beneficial arrangements from port authorities.

The UNCTAD liner shipping connectivity index provides an indicator of a country's position within the global liner shipping network. Liner shipping connectivity is closely related to trade costs and trade competitiveness. Table 2.7 depicts the ranking of selected countries in different regions according to their index in 2018. The liner shipping connectivity index reflects both changes in demand and decisions taken by carriers, which in turn depend on their strategic vessel deployment and responses to port investments and reforms in the container ports of countries (for further analysis of the causes and implications of changes in maritime connectivity, see chapter 6 of the *Review of Maritime Transport 2017*). The following countries experienced a significant increase in the 2018 index compared with the 2017 index: United Arab Emirates, by 179.1 per cent; Maldives, by 124.9 per cent; Mauritania, by 77.1 per cent; Eritrea, by 73.3 per cent; the Federated States of Micronesia, by 69.2 per cent; and Cameroon, by 66.5 per cent. By contrast, the following economies experienced the sharpest decreases in the 2018 index: Ukraine, by 60.6 per cent; Albania, by 48.6 per cent; Montenegro, by 47.6 per cent; New Zealand, by 42.9 per cent; Northern Mariana Islands, by 34.7 per cent; and Yemen, by 31.7 per cent.

**Table 2.6 Number of operators and maximum ship size in selected small island developing States and vulnerable economies, 2017 and 2018**

| | Number of operators | | Maximum ship size, 2018 (20-foot equivalent units) | Maximum ship size change, 2017–2018 (20-foot equivalent units) |
|---|---|---|---|---|
| | 2017 | 2018 | | |
| Martinique | 4 | 3 | 2 626 | - 198 |
| Northern Mariana Islands | 5 | 3 | 1 357 | - 724 |
| Guam | 5 | 4 | 2 692 | — |
| Marshall Islands | 5 | 4 | 1 617 | — |
| Saint Vincent and the Grenadines | 6 | 4 | 1 282 | - 7 |
| Sudan | 9 | 4 | 5 368 | -1 551 |
| Guadeloupe | 6 | 5 | 2 626 | - 198 |
| Somalia | 6 | 5 | 2 394 | - 34 |
| Cuba | 7 | 6 | 2 095 | - 456 |
| Reunion | 7 | 6 | 6 639 | - 311 |

*Source:* UNCTAD secretariat calculations, based on data from MDS Transmodal.
*Note:* Figures based on monthly schedules of liner companies for 1 May 2017 and 1 May 2018.

**Table 2.7 Level of maritime connectivity, 2018**

| | Best connected countries and/or territories | 2018 index | Least connected countries and/or territories | 2018 index |
|---|---|---|---|---|
| Global leaders | 1. China | 187.8 | 1. Norfolk Island | 0.6 |
| | 2. Singapore | 133.9 | 2. Christmas Island | 0.9 |
| | 3. Korea, Rep. | 118.8 | 3. Cayman Islands | 1.2 |
| | 4. Hong Kong (China) | 113.5 | 4. Bermuda | 1.5 |
| | 5. Malaysia | 109.9 | 5. Tuvalu | 1.6 |
| | 6. Netherlands | 98.0 | 6. Wallis and Futuna Islands | 1.6 |
| | 7. Germany | 97.1 | 7. Nauru | 1.9 |
| | 8. United States | 96.7 | 8. Cook Islands | 2.0 |
| | 9. United Kingdom | 95.6 | 9. Greenland | 2.3 |
| | 10. Belgium | 91.1 | 10. Timor-Leste | 2.5 |
| Africa | 1. Morocco | 71.5 | 11. Montserrat | 3.0 |
| | 2. Egypt | 70.3 | 12. Montenegro | 3.0 |
| | 3. South Africa | 40.1 | 13. Albania | 3.0 |
| | 4. Djibouti | 37.0 | 14. Anguilla | 3.2 |
| | 5. Togo | 35.9 | 15. Palau | 3.3 |
| Asia | 1. United Arab Emirates | 83.9 | 16. Federated States of Micronesia | 3.4 |
| | 2. Taiwan, province of China | 78.0 | 17. Antigua and Barbuda | 3.5 |
| | 3. Japan | 76.8 | 18. Democratic Republic of the Congo | 3.5 |
| | 4. Sri Lanka | 72.5 | 19. British Virgin Islands | 3.7 |
| | 5. Vietnam | 68.8 | 20. Saint Kitts and Nevis | 3.7 |
| Latin America and the Caribbean | 1. Panama | 56.6 | 21. United States Virgin Islands | 4.3 |
| | 2. Colombia | 50.1 | 22. Northern Mariana Islands | 4.4 |
| | 3. Mexico | 49.1 | 23. Saint Vincent and the Grenadines | 4.4 |
| | 4. Peru | 43.8 | 24. Saint Lucia | 4.8 |
| | 5. Chile | 42.9 | 25. Kiribati | 4.8 |
| | | | 26. Faroe Islands | 4.8 |
| | | | 27. Dominica | 4.8 |

*Source:* UNCTAD secretariat calculations, based on liner shipping connectivity index.
*Note:* For the liner shipping connectivity index of each country, see http://stats.unctad.org/lsci.
*Abbreviation:* SAR, Special Administrative Region.

## C. SHIP REGISTRATION

Most commercial ships are registered under a flag that differs from the flag of the country of ownership (table 2.3). The three leading flags of registration are those of countries that are not major shipowners, namely Panama, the Marshall Islands and Liberia (table 2.8). The Marshall Islands has continued to increase its market share in recent years and, as at January 2018, had become the world's second largest registry. The fourth and fifth largest registries are Hong Kong (China) and Singapore, and accommodate both owners headquartered in each economy and owners from other economies.

The registries specialize in different vessel types (table 2.9). With regard to commercial value, almost 24 per cent of the world's dry bulk carrier fleet is registered in Panama, including tonnage mostly owned by Japan; 17 per cent of the oil and gas tanker fleet is registered in the Marshall Islands, including many Greece-owned tankers; 27 per cent of the ferry and passenger ship fleet, including United States-owned cruise ships, is registered in the Bahamas; and 16 per cent of the container ship fleet is registered in Liberia, including many Germany-owned ships. As the market share of Germany among the main shipowning countries has declined in recent years, so has the market share of the registries that cater mostly for this market, including Liberia and Antigua and Barbuda, which recorded the greatest decrease in 2017.

**Table 2.8 Top 35 flags of registration by dead-weight tonnage, 2018**

| | Number of vessels | Vessel share of world total (percentage) | Dead-weight tonnage (thousands of tons) | Share of world total dead-weight tonnage (percentage) | Cumulated share of dead-weight tonnage (percentage) | Average vessel size (dead-weight tons) | Dead-weight tonnage change, 2017–2018 (percentage) |
|---|---|---|---|---|---|---|---|
| Panama | 7 914 | 8.40 | 335 888 | 17.46 | 17.46 | 42 442 | -2.04 |
| Marshall Islands | 3 419 | 3.63 | 237 826 | 12.36 | 29.82 | 69 560 | 9.91 |
| Liberia | 3 321 | 3.53 | 223 668 | 11.63 | 41.44 | 67 350 | 3.10 |
| Hong Kong (China) | 2 615 | 2.78 | 181 488 | 9.43 | 50.88 | 69 403 | 4.60 |
| Singapore | 3 526 | 3.74 | 127 880 | 6.65 | 57.52 | 36 268 | 2.93 |
| Malta | 2 205 | 2.34 | 108 759 | 5.65 | 63.18 | 49 324 | 7.45 |
| China | 4 608 | 4.89 | 84 184 | 4.38 | 67.55 | 18 269 | 6.79 |
| Bahamas | 1 418 | 1.51 | 76 659 | 3.98 | 71.54 | 54 061 | -4.14 |
| Greece | 1 343 | 1.43 | 72 345 | 3.76 | 75.30 | 53 868 | 0.14 |
| Japan | 5 299 | 5.63 | 37 536 | 1.95 | 77.25 | 7 084 | 7.88 |
| Cyprus | 1 020 | 1.08 | 34 848 | 1.81 | 79.06 | 34 165 | 3.16 |
| Isle of Man | 412 | 0.44 | 27 275 | 1.42 | 80.48 | 66 201 | 9.15 |
| Indonesia | 9 053 | 9.61 | 22 313 | 1.16 | 81.64 | 2 465 | 9.95 |
| Madeira | 422 | 0.45 | 19 105 | 0.99 | 82.63 | 45 273 | 27.11 |
| India | 1 719 | 1.83 | 18 481 | 0.96 | 83.59 | 10 751 | 6.70 |
| Danish International Register of Shipping | 452 | 0.48 | 18 165 | 0.94 | 84.53 | 40 188 | 7.80 |
| Norwegian International Ship Register | 519 | 0.55 | 18 056 | 0.94 | 85.47 | 34 790 | -0.76 |
| United Kingdom | 1 157 | 1.23 | 16 764 | 0.87 | 86.34 | 14 489 | 5.79 |
| Italy | 1 405 | 1.49 | 15 090 | 0.78 | 87.13 | 10 740 | -5.54 |
| Republic of Korea | 1 897 | 2.01 | 14 426 | 0.75 | 87.88 | 7 605 | -4.89 |
| Saudi Arabia | 380 | 0.40 | 13 522 | 0.70 | 88.58 | 35 584 | 238.90 |
| United States | 3 692 | 3.92 | 12 045 | 0.63 | 89.21 | 3 262 | 2.48 |
| Bermuda | 160 | 0.17 | 10 612 | 0.55 | 89.76 | 66 325 | -3.01 |
| Malaysia | 1 704 | 1.81 | 10 230 | 0.53 | 90.29 | 6 004 | 3.88 |
| Germany | 629 | 0.67 | 9 936 | 0.52 | 90.81 | 15 797 | -5.51 |
| Russian Federation | 2 625 | 2.79 | 8 613 | 0.45 | 91.25 | 3 281 | 3.45 |
| Antigua and Barbuda | 853 | 0.91 | 8 578 | 0.45 | 91.70 | 10 056 | -15.02 |
| Belgium | 192 | 0.20 | 8 497 | 0.44 | 92.14 | 44 255 | 5.87 |
| Viet Nam | 1 863 | 1.98 | 8 176 | 0.42 | 92.57 | 4 389 | 2.01 |
| Turkey | 1 263 | 1.34 | 7 740 | 0.40 | 92.97 | 6 128 | -3.48 |
| Netherlands | 1 233 | 1.31 | 7 326 | 0.38 | 93.35 | 5 942 | -0.83 |
| Thailand | 807 | 0.86 | 6 212 | 0.32 | 93.67 | 7 698 | 15.21 |
| Cayman Islands | 165 | 0.18 | 6 155 | 0.32 | 93.99 | 37 303 | 10.17 |
| Philippines | 1 615 | 1.72 | 5 683 | 0.30 | 94.29 | 3 519 | -8.41 |
| French Flag Register | 94 | 0.10 | 5 031 | 0.26 | 94.55 | 53 521 | -4.68 |
| **Total, top 35 flags** | **70 999** | **75.40** | **1 819 112** | **94.55** | **94.55** | **25 622** | **-** |
| *Rest of world* | *23 170* | *24.60* | *104 890* | *5.45* | *5.45* | *4 527* | - |
| **World total** | **94 169** | **100.00** | **1 924 002** | **100.00** | **100.00** | **20 431** | **3.34** |

*Source:* UNCTAD secretariat calculations, based on data from Clarksons Research.
*Notes:* Propelled seagoing merchant vessels of 100 gross tons and above, as at 1 January. For a complete listing of countries, see http://stats.unctad.org/fleet.
*Abbreviation:* SAR, Special Administrative Region.

**Table 2.9 Leading flags of registration by value of principal vessel type, 2018**
(Millions of dollars)

| | Oil tankers | Dry bulk carriers | General cargo ships | Container ships | Gas carriers | Chemical tankers | Offshore vessels | Ferries and passenger ships | Other | Total |
|---|---|---|---|---|---|---|---|---|---|---|
| Panama | 12 564 | 46 799 | 3 909 | 13 601 | 8 027 | 5 286 | 20 889 | 9 920 | 7 506 | 128 501 |
| Marshall Islands | 22 479 | 28 088 | 504 | 6 473 | 13 604 | 4 881 | 24 667 | 1 316 | 2 456 | 104 469 |
| Bahamas | 7 430 | 5 042 | 174 | 413 | 9 885 | 140 | 26 807 | 26 911 | 2 747 | 79 551 |
| Liberia | 15 284 | 21 158 | 1 039 | 16 388 | 4 548 | 2 045 | 11 022 | 151 | 1 648 | 73 281 |
| Hong Kong (China) | 9 370 | 24 785 | 1 968 | 14 983 | 3 589 | 1 982 | 324 | 50 | 122 | 57 173 |
| Singapore | 10 764 | 13 346 | 1 188 | 10 686 | 5 011 | 2 799 | 7 617 | — | 1 778 | 53 189 |
| Malta | 8 769 | 11 684 | 1 815 | 7 911 | 4 106 | 2 246 | 4 977 | 10 045 | 594 | 52 148 |
| China | 4 900 | 13 811 | 2 583 | 2 568 | 915 | 1 557 | 7 192 | 4 693 | 2 304 | 40 523 |
| Italy | 1 400 | 1 113 | 2 772 | 121 | 298 | 550 | 608 | 12 044 | 354 | 19 260 |
| Greece | 8 832 | 3 935 | 187 | 237 | 4 364 | 63 | 1 | 1 447 | 100 | 19 166 |
| United Kingdom | 562 | 661 | 1 145 | 3 765 | 447 | 723 | 4 727 | 4 315 | 496 | 16 840 |
| Bermuda | 413 | 173 | 9 | 86 | 6 412 | 336 | 2 295 | 6 466 | — | 16 191 |
| Japan | 2 417 | 3 718 | 1 926 | 425 | 1 551 | 157 | 582 | 2 905 | 1 895 | 15 575 |
| Cyprus | 721 | 5 396 | 850 | 1 769 | 861 | 306 | 2 071 | 616 | 843 | 13 433 |
| Norwegian International Ship Register | 1 672 | 1 860 | 239 | — | 2 729 | 1 031 | 3 372 | 697 | 1 230 | 12 831 |
| Isle of Man | 2 646 | 2 638 | 267 | 268 | 2 545 | 337 | 3 358 | 26 | 16 | 12 101 |
| Netherlands | 136 | 161 | 3 675 | 208 | 482 | 173 | 1 615 | 3 307 | 1 018 | 10 776 |
| Norway | 269 | 109 | 150 | — | 101 | 148 | 7 227 | 1 865 | 2 | 9 871 |
| Danish International Register of Shipping | 1 082 | 81 | 533 | 5 783 | 819 | 559 | 468 | 431 | 105 | 9 861 |
| Indonesia | 1 580 | 725 | 1 580 | 677 | 542 | 317 | 2 276 | 1399 | 36 | 9 132 |
| United States | 1 311 | 36 | 528 | 629 | — | 33 | 3 727 | 1 668 | 721 | 8 654 |
| Malaysia | 673 | 176 | 79 | 67 | 1 837 | 219 | 5 112 | 14 | 133 | 8 310 |
| Madeira | 169 | 1 678 | 362 | 4 292 | 26 | 230 | 1 | 38 | 208 | 7 004 |
| India | 1 580 | 1 079 | 561 | 127 | 230 | 87 | 961 | 293 | 233 | 5 150 |
| Nigeria | 146 | — | 5 | — | — | 80 | 4 905 | 2 | 2 | 5 140 |
| **Subtotal, top 25 flags** | **117 168** | **188 252** | **28 047** | **91 477** | **72 932** | **26 283** | **146 804** | **90 618** | **26 548** | **788 129** |
| *Other* | *13 486* | *10 099* | *15 354* | *8 902* | *5 512* | *7 022* | *28 637* | *11 119* | *5 208* | *105 337* |
| **World total** | **130 654** | **198 351** | **43 401** | **100 379** | **78 443** | **33 305** | **175 440** | **101 737** | **31 756** | **893 467** |

*Source:* UNCTAD secretariat calculations, based on data from Clarksons Research.
*Notes:* Propelled seagoing merchant vessels of 1,000 gross tons and above, as at 1 January.
*Abbreviation:* SAR, Special Administrative Region.

**Table 2.10 Distribution of dead-weight tonnage capacity of vessel types by country group of registration, 2018**
(Percentage)

| | Total fleet | Oil tankers | Dry bulk carriers | General cargo ships | Container ships | Other |
|---|---|---|---|---|---|---|
| Developed countries | 23.14 | 25.21 | 18.66 | 27.87 | 29.02 | 26.24 |
| | *0.23* | *0.67* | *-0.10* | *0.00* | *0.48* | *0.12* |
| Countries with economies in transition | 0.67 | 0.88 | 0.19 | 5.54 | 0.05 | 1.06 |
| | *-0.01* | *-0.04* | *-0.01* | *0.15* | *0.00* | 0.02 |
| Developing countries | 75.94 | 73.81 | 81.13 | 65.20 | 70.85 | 71.43 |
| | *-0.18* | *-0.51* | *0.13* | *-0.23* | *-0.31* | *-0.33* |
| Of which: | | | | | | |
| Africa | 12.49 | 13.87 | 11.23 | 6.98 | 18.17 | 8.91 |
| | *-0.07* | *-1.40* | *0.77* | *0.44* | *-0.36* | *-0.30* |
| America | 23.47 | 19.63 | 27.27 | 20.37 | 16.44 | 28.30 |
| | *-1.35* | *-1.40* | *-1.58* | *-0.31* | *-1.47* | *-0.50* |
| Asia | 27.21 | 24.45 | 28.91 | 35.01 | 30.45 | 21.53 |
| | *0.53* | *1.33* | *-0.10* | *0.15* | *1.14* | *0.54* |
| Oceania | 12.76 | 2.84 | 13.72 | 2.84 | 5.78 | 12.69 |
| | *0.71* | *0.75* | *1.03* | *-0.52* | *0.39* | *-0.07* |
| Unknown and other | 0.25 | 0.10 | 0.03 | 1.38 | 0.09 | 1.27 |
| | *-0.04* | *-0.12* | *-0.01* | *0.08* | *-0.18* | *0.19* |
| **World total** | **100.00** | **100.00** | **100.00** | **100.00** | **100.00** | **100.00** |

*Source:* UNCTAD secretariat calculations, based on data from Clarksons Research.
*Notes:* Propelled seagoing merchant vessels of 100 gross tons and above, as at 1 January. Annual change in italics.

The major open registries are hosted by developing countries. Accordingly, developing countries account for almost 76 per cent of the global national flag tonnage, developed countries account for 23 per cent and countries with economies in transition account for less than 1 per cent (table 2.10).

## D. SHIPBUILDING, DEMOLITION AND NEW ORDERS

### 1. Delivery of newbuildings

In 2017, total delivery amounted to 65 million gross tons, equivalent to 5.2 per cent of the start-of-year fleet in 2017 (table 2.11). In addition in 2017, 23 million gross tons were scrapped, leading to a net growth in the world fleet of 42 million gross tons, equivalent to a growth rate of 3.3 per cent.

The dry bulk sector saw the largest tonnage of newbuilding entering the fleet, with more than 20 million gross tons reported delivered; this sector also saw the highest level of scrapping activity, at more than 8 million gross tons, leading to a net growth in the dry bulk fleet of 2.9 per cent. Oil tankers saw less newbuilding activity but also less scrapping, resulting in greater net growth in the fleet, at almost 5 per cent. General cargo ships recorded more scrapping than newbuildings, leading to a negative growth rate in this sector. The largest shipbuilding countries continued to be China, the Republic of Korea and Japan, which together accounted for 90.5 per cent of gross tons delivered in 2017. China has the largest market shares in dry bulk carriers and general cargo ships. The Republic of Korea is strongest in oil tankers, container ships and gas carriers. Japan has its largest market share in chemical tankers and bulk carriers. The rest of the world, comprising mostly countries in Europe, is strongest in offshore vessels and passenger ships, including cruise ships.

### 2. Ship demolition

Ship demolitions in 2017 were almost one quarter less in gross tons than in 2016, an indicator of improved market optimism. Bulk carrier and container ship scrapping slowed in line with improved market conditions but tanker recycling increased. The most ship scrapping continued to take place in India, followed by Bangladesh and Pakistan (table 2.12).

**Table 2.11 Deliveries of newbuildings by major vessel type and countries of construction, 2017**
(Thousands of gross tons)

| | China | Republic of Korea | Japan | Philippines | Rest of world | Total |
|---|---|---|---|---|---|---|
| Oil tankers | 5 330 | 10 859 | 1 835 | 472 | 1 213 | 19 709 |
| Dry bulk carriers | 11 982 | 640 | 7 713 | 480 | 236 | 21 052 |
| General cargo ships | 588 | 75 | 186 | — | 233 | 1 082 |
| Container ships | 3 105 | 5 873 | 1 408 | 974 | 451 | 11 813 |
| Gas carriers | 708 | 3 973 | 439 | 52 | 12 | 5 185 |
| Chemical tankers | 654 | 6 | 531 | — | 137 | 1 329 |
| Offshore vessels | 409 | 473 | 145 | 0 | 647 | 1 675 |
| Ferries and passenger ships | 166 | — | 197 | 1 | 1 174 | 1 537 |
| Other | 395 | 609 | 482 | — | 121 | 1 607 |
| **Total** | **23 339** | **22 509** | **12 937** | **1 980** | **4 224** | **64 989** |

*Source:* UNCTAD secretariat calculations, based on data from Clarksons Research.
*Notes:* Propelled seagoing merchant vessels of 100 gross tons and above. For more detailed data on other shipbuilding countries, see http://stats.unctad.org/shipbuilding.

**Table 2.12 Reported tonnage sold for demolition by major vessel type and country of demolition, 2017**
(Thousands of gross tons)

| | India | Bangladesh | Pakistan | China | Unknown – Indian subcontinent | Turkey | Other/unknown | World total |
|---|---|---|---|---|---|---|---|---|
| Oil tankers | 1 935 | 3 245 | 0 | 1 | 749 | 12 | 40 | 5 982 |
| Dry bulk carriers | 1 062 | 1 460 | 2 527 | 2 464 | 470 | 139 | 0 | 8 123 |
| General cargo ships | 420 | 155 | 102 | 82 | 0 | 312 | 108 | 1 178 |
| Container ships | 1 755 | 892 | 748 | 650 | 140 | 309 | 3 | 4 498 |
| Gas carriers | 145 | 59 | 0 | 4 | 0 | 173 | 5 | 387 |
| Chemical tankers | 109 | 35 | 0 | 2 | 44 | 0 | 6 | 196 |
| Offshore vessels | 318 | 57 | 77 | 90 | 157 | 128 | 404 | 1 230 |
| Ferries and passenger ships | 165 | 35 | 5 | 0 | 0 | 51 | 21 | 277 |
| Other | 415 | 321 | 0 | 152 | 0 | 133 | 23 | 1 044 |
| **Total** | **6 323** | **6 260** | **3 459** | **3 445** | **1 560** | **1 257** | **611** | **22 916** |

*Source:* UNCTAD secretariat calculations, based on data from Clarksons Research.
*Notes:* Propelled seagoing merchant vessels of 100 gross tons and above. Estimates for all countries are available at http://stats.unctad.org/shipscrapping.

Figure 2.7 World tonnage on order, 2000–2018
(Thousands of dead-weight tons)

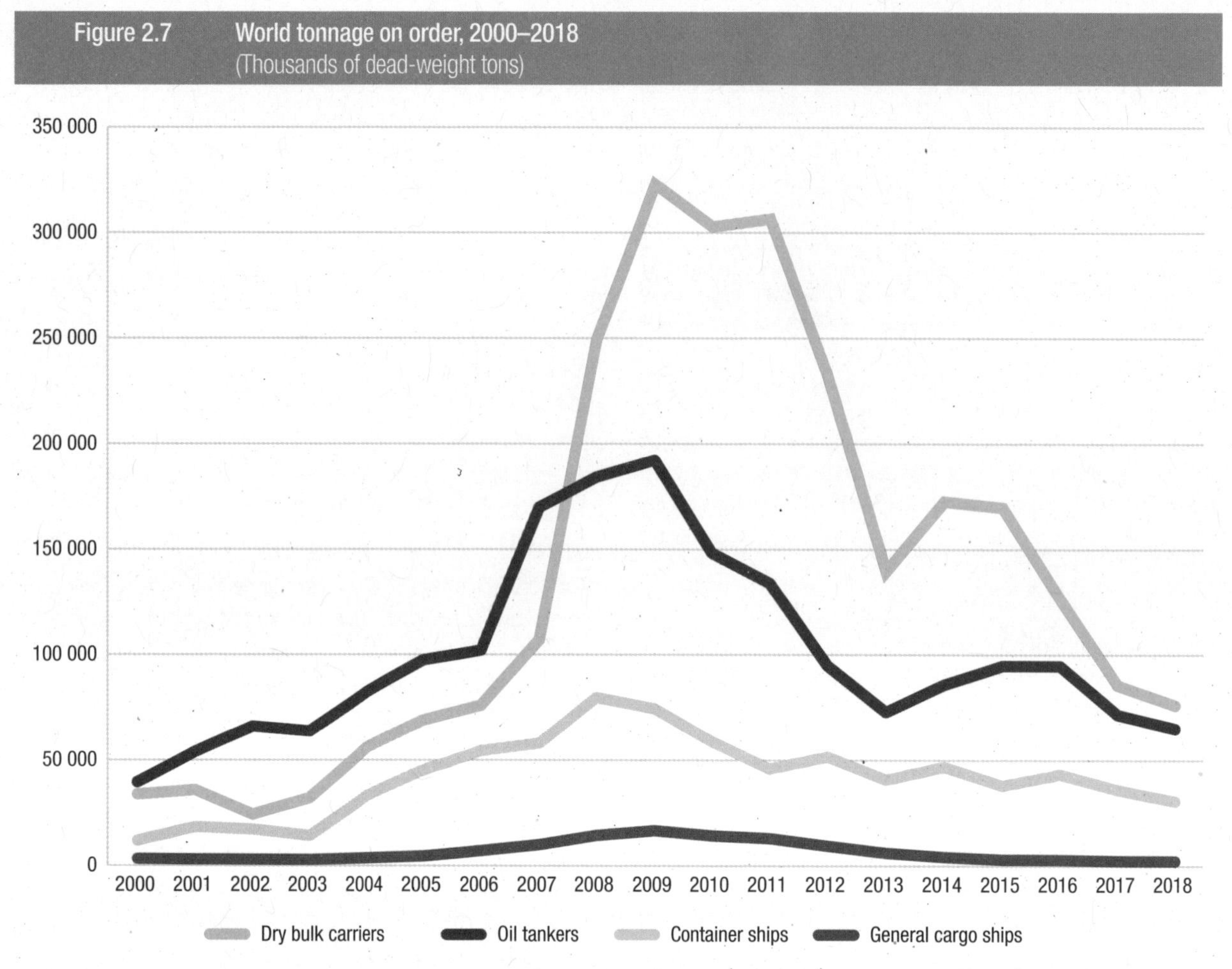


*Source:* UNCTAD secretariat calculations, based on data from Clarksons Research.
*Notes:* Propelled seagoing merchant vessels of 100 gross tons and above, as at 1 January.

## 3. Tonnage on order

The tonnage on order for all main vessel types further decreased between 2017 and 2018 (figure 2.7). Compared with the peaks in 2008 and 2009, the current tonnage on order has decreased by 62 per cent for container ships, 66 per cent for oil tankers, 76 per cent for dry bulk carriers and 85 per cent for general cargo ships. With regard to TEUs, two thirds of the container ship orderbook is for ships of 14,000 TEUs and above.

With regard to shipbuilding countries, China accounts for 41.6 per cent of the dwt on order, followed by the Republic of Korea at 24.3 per cent and Japan at 23.6 per cent (figure 2.8). Nearly all shipbuilding of cargo-carrying vessels takes place in Asia. The other shipbuilding countries in the figure focus on passenger ships and specialized ships such as offshore vessels.

# E. ASSESSING GENDER EQUALITY ASPECTS IN SHIPPING

An increasing number of women are entering the shipping industry in all roles, including seafaring and operations, chartering, insurance and law. More women are also enrolling in maritime-related studies. This may be attributed to efforts to advance the role of women in the maritime industry, including through IMO initiatives in global capacity-building and International Labour Organization and International Transport Workers' Federation initiatives in standard-setting.

Challenges remain, however. The level of women's participation in the maritime industry remains low, at an estimated 2 per cent, and patterns of job segregation exist (World Economic Forum, 2015). According to Maritime HR Association survey data from 2017, women who work in the shipping industry are paid on average 45 per cent less than men and fill solely 7 per cent of management positions (HR Consulting, 2017). Table 2.13 depicts three outcomes of the lack of gender equality in the maritime industry.

Overcoming the lack of gender equality in the maritime industry may be a core element in addressing the shortage of skilled professionals in the sector, which could impact shipping operations in the future. Two main factors help explain the low level of participation of women in the transport sector, namely working conditions and gender stereotyping (Turnbull, 2013).

With regard to seafaring roles, working conditions refer, for example, to a lack of amenities on ships and to alternatives for accommodating interruptions that may occur due to

**Figure 2.8 Tonnage on order by shipbuilding country, 2018**

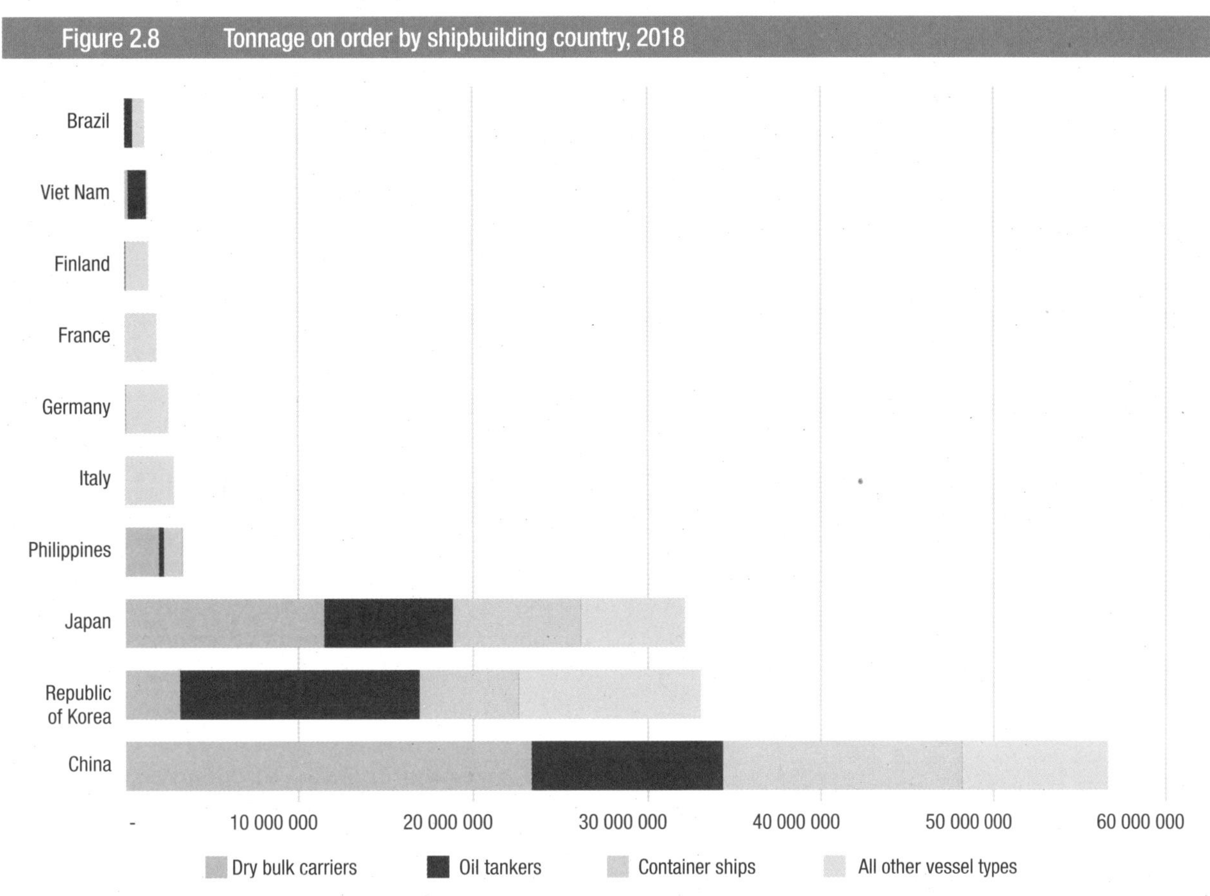


*Source:* UNCTAD secretariat calculations, based on data from Clarksons Research.
*Notes:* Propelled seagoing merchant vessels of 100 gross tons and above, as at 1 January.

**Table 2.13 Lack of gender equality in the maritime industry**

| | |
|---|---|
| 1. Levels of seniority | Over 76 per cent of the women's workforce operates at administrative, junior and professional level roles, with few reaching managerial levels or higher |
| | Only 0.17 per cent of women have places on executive leadership teams |
| | The greatest challenge for women appears to be progressing from a professional to a senior professional level |
| 2. Job functions | In technical, marine, safety and quality-related functions, women represent 14 per cent of the workforce, likely linked to the low number of women seafarers moving to onshore positions. Women employees are heavily weighted at the junior level and 90 per cent of all other employees are men, suggesting that there are currently few opportunities for women to progress in such functions |
| | In chartering functions, women represent 17 per cent of the workforce. Although the majority remain at the administrative and junior levels, there is better representation at the professional, senior professional and managerial levels than in the previous category |
| | In commercial functions, women represent 33 per cent of the workforce, with better representation at all levels than in the other categories |
| 3. Salaries | The difference in the average salary of men and women is 45 per cent |
| | Countries with the greatest salary differences do not employ any women on executive leadership teams and employ few at the directorial level |
| | Except at the junior and administrative levels, men are paid on average more than women |

*Source:* HR Consulting, 2017.
*Note:* The survey reflects data for worldwide onshore positions in organizations members of the Maritime HR Association.

childbearing and other responsibilities of care, such as through the provision of flexible working hours, maternity benefits and childcare facilities. Working conditions can also refer to exposure to harassment and violence, a recurrent concern expressed in the seafaring sector (MacNeil and Ghosh, 2016). Such elements lead to a lack of interest in pursuing a career in the maritime sector or to early departures from maritime industry careers. A study on the career awareness of cadets in South Africa showed that the expected span of careers at sea among women was 10 years and that many contemplated leaving their positions during their early 30s (Ruggunan and Kanengoni, 2017).

Gender stereotyping, that is, a cultural perception that women are less able to meet the demands of a career in this sector, is present with regard to physical roles in seafaring operations, as well as in other segments of the maritime industry, such as insurance and law, which can lead to workplaces that are unwelcoming or openly hostile towards women (Wu et al., 2017). Gender stereotyping also encompasses inappropriate sexual comments, persistent sexual invitations, unwanted physical contact and bullying (MacNeil and Ghosh, 2016; Turnbull, 2013). In addition, it includes discriminatory practices, in particular in lower ranks and in the younger age demographic (Ship Technology, 2017). With regard to onshore managerial roles, a study on women's maritime careers in Eastern and Southern Africa showed that gender stereotyping was closely related to the work-intensive pattern of the professional progression of women, aimed at achieving success in the "man-made" system of the maritime industry, because women perceived that they had to devote extra time and energy compared with men peers in order to achieve similar results, due to the distrust of employers with regard to their competence and ability to perform as maritime professionals and to a lack of recognition of their contributions (Bhirugnath-Bhookhum and Kitada, 2017).

Working conditions and gender stereotyping are closely linked. For example, to fit in in men-dominated environments in the seafaring profession, women may adopt behaviours suggestive of masking perceived feminine attributes and emphasizing masculinity, such as with regard to dress and socialization with peers (Acejo and Abila, 2016). Efforts to integrate women into the seafaring profession and erase gender differentials have been both ambivalent and contradictory, and may conversely reinforce gender biases against the participation of women in the workplace (Acejo and Abila, 2016). For example, some shipping companies require prior seafaring experience to access managerial roles, in a context in which companies are often reluctant to take on women cadets, resulting in an unequal playing field with regard to onshore career progression.

Several international voluntary frameworks and programmes have been put in place at the international and regional levels to meet different aspects of these challenges. For example, in 1989, IMO launched the Women in Development Programme to enhance the capabilities of women in the sector; this programme is now entitled Programme on the Integration of Women in the Maritime Sector, and its main objective is to facilitate access to high-level technical training for women maritime officials. In addition, the International Transport Workers' Federation has instituted a code of conduct on eliminating shipboard harassment and bullying. With regard to factors affecting professional progression in onshore roles, frameworks have been prepared by IMO, regional organizations and women's associations. However, their implementation differs significantly at the national level. For example, Kenya, Mauritius, Seychelles and South Africa have developed practices aimed at empowering women in managerial positions and at retaining women employees, including through the use of flexible working hours (Bhirugnath-Bhookhum and Kitada, 2017).

Overcoming such causes of the lack of gender equality in the maritime industry is likely to require coordinated efforts by several stakeholders, including shipping companies, crewing agencies, freight companies, trade unions and seafarers' welfare organizations. Measures could encompass actions at three levels.

## Educational level

### *Increase awareness of gender equity in maritime academic, operational and business spheres*

Increased awareness is required to promote a more systematic gender-sensitive approach in the profession. This could be achieved, for example, by adding related topics to the curricula of maritime educational institutions and ensuring staff induction and consistent sensitization training at the management, human resources, ship manager and ship master levels, which emphasize issues such as improving on-board conditions and policies to report and address sexual harassment and discrimination.

### *Ensure that training institution curricula are structured to allow graduates to work both onshore and offshore*

Such curricula would allow for career paths that are versatile and for flexibility and the retention of trained, experienced individuals who may not be in a position to work on board vessels.

## Organizational level

### *Ensure adequate maternity benefits and flexibility schemes*

This would facilitate the shift from offshore to onshore positions without penalization in climbing managerial ladders and could contribute to improving the retention of women in the industry.

### *Develop gender-neutral working practices*

Such practices, particularly those focused on hiring and promotion, would help increase the participation of women in the industry at all levels.

## Institutional and national levels

### *Promote the adoption of internationally agreed codes of conduct and standards*

Such codes include the Maritime Labour Convention, 2006, and the International Transport Workers' Federation code of conduct on eliminating shipboard harassment and bullying. Social partners should be involved in the monitoring of enforcement. The creation and adoption of business policies on harassment and bullying, as well as on reporting measures to eliminate such actions, should be encouraged.

***Strengthen and consolidate regional networks***

This would help support the dissemination of best practices as a basis for mainstreaming better gender-related practices in the maritime industry.

***Enhance partnerships between individual institutions and industry association organizations***

Such organizations include the Women's International Shipping and Trading Association. Enhanced partnerships should provide long-term coaching, networking and fellowship opportunities and could contribute to retention, creating further opportunities to advance careers, cooperate, share best practices and work across borders.

***Inspire and empower new generations by identifying women role models in the sector***

This could include the organization of workshops to exchange experiences and the creation of mentoring programmes.

## F. OUTLOOK AND POLICY CONSIDERATIONS

In 2017, with positive developments in demand and freight rates, the world fleet grew slightly faster than in 2016. Yet the industry refrained from an expansion that would have added more capacity than needed, and 2017 was the first year since 2003 for which UNCTAD recorded a lower growth rate for world tonnage than for seaborne trade. However, there are signs that the fleet will expand at a higher rate in 2018 and 2019. With regard to container ships, there has been almost no scrapping in the first half of 2018, and total TEU capacity growth is forecasted to reach 5 per cent by January 2019 (Clarksons Research, 2018). In the medium term, for example, the Republic of Korea aims to build 200 new container and dry cargo ships and establish a maritime industry promotion agency to support the placement of orders for new ships through investments or by guaranteeing the ship purchase programme (Marine Log, 2018). As countries try to support their maritime industries, notably in shipowning and construction, they may effectively subsidize the shipping industry and, indirectly, global trade. If the additional carrying capacity outstrips demand, the resulting surplus capacity will put further pressure on freight rates and thus may create further imbalances. Promoting the construction and operation of new and more efficient vessels should be accompanied by strong scrapping and demolition incentives to manage supply-side capacity.

The recent mergers and continued consolidation in container shipping suggest that an ever lower number of carriers, cooperating in only three major global alliances, will control the supply of shipping services in coming years. From the supply-side perspective, the operational gains due to alliances have effectively added surplus capacity to the market. As cooperation and vessel sharing help to improve capacity utilization, fewer ships are needed for the same cargo volumes and when no-longer-needed ships are not scrapped – and they are not – the resulting surplus puts further downward pressures on freight rates. Policymakers and regulators will need to ensure that members of shipping alliances continue to compete with regard to prices, so that efficiency gains on the supply side may be passed on to shippers in the form of lower freight rates.

A challenge arises if traffic volumes are too low to economically allow for more than a small number of competing carriers. UNCTAD records show a decreasing number of carriers, in particular for services to small island developing States and some vulnerable economies. In such situations, government interventions may be justified, yet in practice may do more harm than good. Assessing the implications of horizontal and vertical integration in the industry and addressing potential negative effects through solutions acceptable to all parties will require the engagement of competition authorities, carriers, shippers and ports. The United Nations Set of Multilaterally Agreed Equitable Rules and Principles for the Control of Restrictive Business Practices provides for consultations between member States in this area.

Average vessel sizes and the fleet of gearless container ships continue to grow. This has important repercussions for investments in terminals to provide the adequate space, infrastructure and equipment needed to service these fleets. As the fleet of geared ships further diminishes, policymakers and port planners need to seize every opportunity to invest in the most appropriate specialized terminals.

An increasing number of women are entering the shipping industry, yet a lack of gender equality remains with regard to levels of seniority, job functions and salaries. Overcoming this gender imbalance in the maritime industry may be a core element in dealing with the shortage of skilled professionals in the sector, which could impact shipping operations in future. In order to address the shortage, two main factors need to be addressed, namely working conditions and gender stereotyping. Efforts need to be made by the industry and by policymakers, and should include coordination between several stakeholders, awareness raising, promotion of the adoption of internationally agreed codes of conduct, revised curricula in training institutions, flexibility schemes and instruments to improve rates of retention and to advance careers.

The supply of shipping services will need to go beyond simply management of vessel operations. The digital transformation of shipping entails a number of opportunities. New technologies include automated navigation and cargo-tracking systems, as well as digital platforms that facilitate operations, trade and the exchange of data. They can potentially reduce costs, facilitate interactions between different actors and raise the maritime supply chain to the next level. Combining on-board systems and digital platforms allows vessels and cargo to become a part of the Internet of things. A key challenge for policymakers is to establish interoperability, so that data can be exchanged seamlessly, at the same time ensuring cybersecurity and the protection of commercially sensitive and private data.

# REFERENCES

Acejo IL and Abila SS (2016). Rubbing out gender: Women and merchant ships. *Journal of Organizational Ethnography*. 5(2):123–138.

Allianz Global Corporate and Specialty (2017). *Safety and Shipping Review 2017*. Munich.

Bhirugnath-Bhookhum M and Kitada M (2017). Lost in success: Women's maritime careers in Eastern and Southern Africa. *Palgrave Communications*. Springer Nature.

Clarksons Research (2018). *Container Intelligence Monthly*. Volume 20. No. 5. May.

Dynamar BV (2018a). *Dynaliners Weekly*. 15 June.

Dynamar BV (2018b). *Dynaliners Weekly*. 13 April.

HR Consulting (2017). *Maritime HR Association: 2017 Market Analysis Report – Gender Diversity in Maritime*. Spinnaker Global.

Lehmacher W (2017). *The Global Supply Chain: How Technology and Circular Thinking Transform Our Future*. Springer International Publishing AG. Cham, Switzerland.

MacNeil A and Ghosh S (2016). Gender imbalance in the maritime industry: Impediments, initiatives and recommendations. *Australian Journal of Maritime and Ocean Affairs*. 9(1):42–55.

*Marine Log* (2018). [Republic of] Korea unveils restructuring plan for shipping and shipyards. 5 April.

Right Ship (2018). Where are the most efficient vessels built? Available at https://site.rightship.com/about-rightship/insights/.

Ruggunan S and Kanengoni H (2017). Pursuing a career at sea: An empirical profile of South African cadets and implications for career awareness. *Maritime Policy and Management*. 44(3):289–303.

Ship Technology (2017). Women in shipping: Pushing for gender diversity. 23 August.

Turnbull P (2013). Promoting the employment [of] women in the transport sector: Obstacles and policy options. Working Paper No. 298. International Labour Organization.

World Economic Forum (2015). Why we need more women in maritime industries. 4 September.

Wu C-L, Chen S-Y, Ye K-D and Ho Y-W (2017). Career development for women in [the] maritime industry: Organization and socialization perspectives. *Maritime Policy and Management*. 44(7):882–898.

# ENDNOTES

1. Data in this chapter concerning tonnage and number of ships in the world fleet was provided by Clarksons Research. Unless stated otherwise, the vessels covered in the UNCTAD analysis include all propelled seagoing merchant vessels of 100 gross tons and above, including offshore drillships and floating production, storage and offloading units. Military vessels, yachts, waterway vessels, fishing vessels and offshore fixed and mobile platforms and barges are not included. Data on fleet ownership only cover ships of 1,000 gross tons and above, as information on the true ownership of smaller ships is often not available. For more detailed data on the world fleet, including registration, ownership, building and demolition, as well as other maritime statistics, see http://stats.unctad.org/maritime.
2. The aggregate fleet values published by Clarksons Research are calculated from estimates of the value of each vessel based on type, size and age. Values are estimated for all oil/product tankers, bulk carriers, combined carriers, container ships and gas carriers with reference to matrices based on representative newbuilding, second-hand and demolition values provided by Clarksons Platou brokers. For other vessel types, values are estimated with reference to individual valuations, recently reported sales and residual values calculated from reported newbuilding prices. As coverage concerning specialized and non-cargo vessels may not be complete, figures might not accurately represent the total value of the world merchant fleet above 100 gross tons. Desktop estimates are made on the basis of prompt charter-free delivery, as between a willing buyer and a willing seller for cash payment under normal commercial terms. For the purposes of this exercise, all vessels are assumed to be in good and seaworthy condition.
3. For further discussion on this issue, see the documentation considered at the seventeenth session of the Intergovernmental Group of Experts on Competition Law and Policy, held from 11 to 13 July 2018, available at http://unctad.org/en/pages/MeetingDetails.aspx?meetingid=1675; the article on consolidation in liner shipping in UNCTAD Transport and Trade Facilitation Newsletter No. 76; and chapter 6 of the *Review of Maritime Transport 2017*. The liner shipping connectivity index, liner shipping bilateral connectivity index and information on calculations for the indices are available at http://stats.unctad.org/maritime.

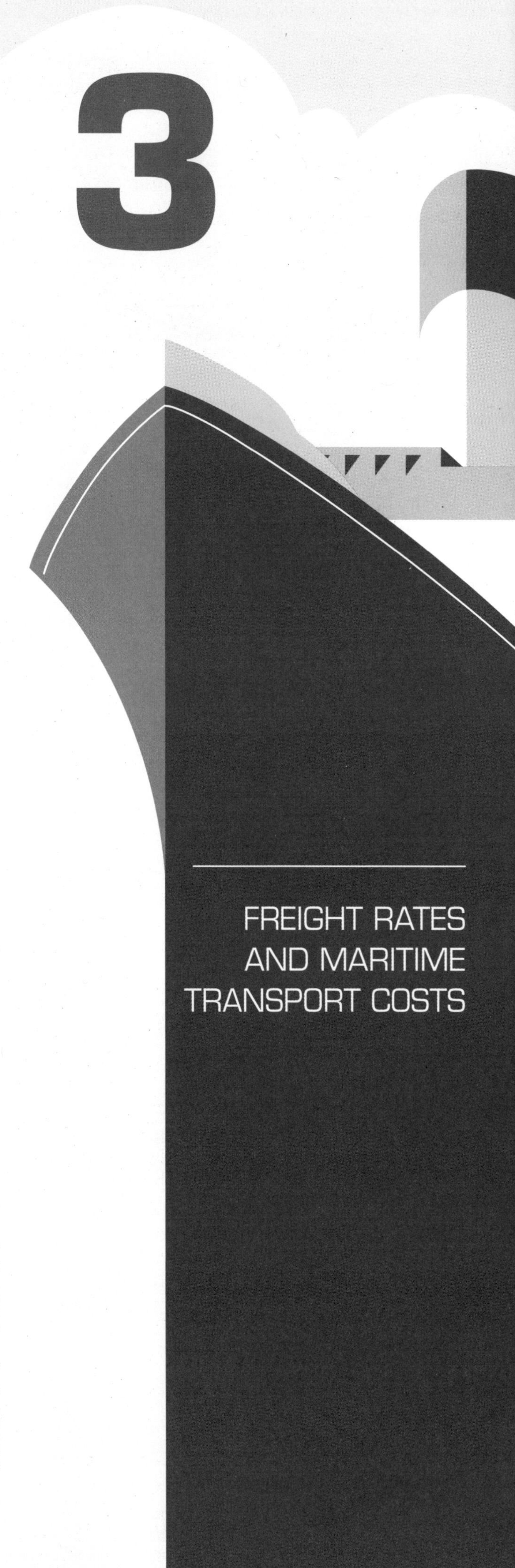


In 2017 and early 2018, the global shipping industry saw a marked improvement of fundamentals in most market segments, with the exception of the tanker market. Key drivers were the combined strengthening in global demand, on the one hand, and the reduced fleet growth, on the other. Overall, freight rates improved across all markets in 2017, with the exception of tankers.

Container freight rate levels increased, and averages surpassed performance in 2016. A better supply–demand balance in container ship markets, underpinned by stronger demand, was the main driver. The container shipping industry ended 2017 with a total profit of roughly $7 billion, driven mainly by a significant increase in transported volumes, freight rates and revenue, as well as proactive operational management discipline.

During the year, consolidation, whether in the form of alliances or mergers and acquisitions, persevered in the container industry in response to the negative environment that the industry has been facing in recent years. While outright negative impacts on trade and costs have not been reported, there are remaining concerns about the impact of growing market concentration on competition and the level playing field. Competition authorities and regulators, as well as transport analysts and international entities such as UNCTAD, should therefore remain vigilant. In this respect, the seventeenth session of the Intergovernmental Group of Experts on Competition Law and Policy held in Geneva in July 2018, provided a timely opportunity to bring together competition authority representatives and other stakeholders from the sector to reflect upon some of these concerns and assess their extent and potential implications for shipping and seaborne trade, as well as the role of competition law and policy in addressing these concerns. Delegates called upon UNCTAD to continue its analytical work in the area of international maritime transport, including the monitoring and analysis of the effects of cooperative arrangements and mergers not only on freight rates but also on the frequency, efficiency, reliability and quality of shipping services.

In 2017, the bulk freight market recorded a remarkable surge, which translated into clear gains for carriers, thereby compensating the depressed earnings of 2016. The improvement was largely driven by faster growth in seaborne dry bulk trade and moderate growth in supply. The tanker market was under pressure in 2017.

A key development is the current debate at IMO regarding the introduction of a set of short- to long-term measures to help curb carbon emissions from international shipping. Depending on the outcome of relevant negotiations and the specific design of any future instrument to be adopted, it will be important to assess the related potential implications for carriers, shippers, operating and transport costs, as well as costs for trade. It will also be important to consider the gains and benefits that may derive from these measures, including market-based instruments in shipping and how these could be directed to address the needs of developing countries, especially in terms of their transport cost burden and their ability to access the global marketplace. Some of the main developments at IMO to address greenhouse gas emissions from ships and issues, namely in the context of market-based instruments, are considered in this chapter.

# FREIGHT MARKETS 2017

In 2017, **freight rates improved across all markets**, with the exception of tankers.

## CONTAINER FREIGHT RATES

A better supply–demand balance in container-ship markets, underpinned by stronger demand, was the main driver for improved freight rates.

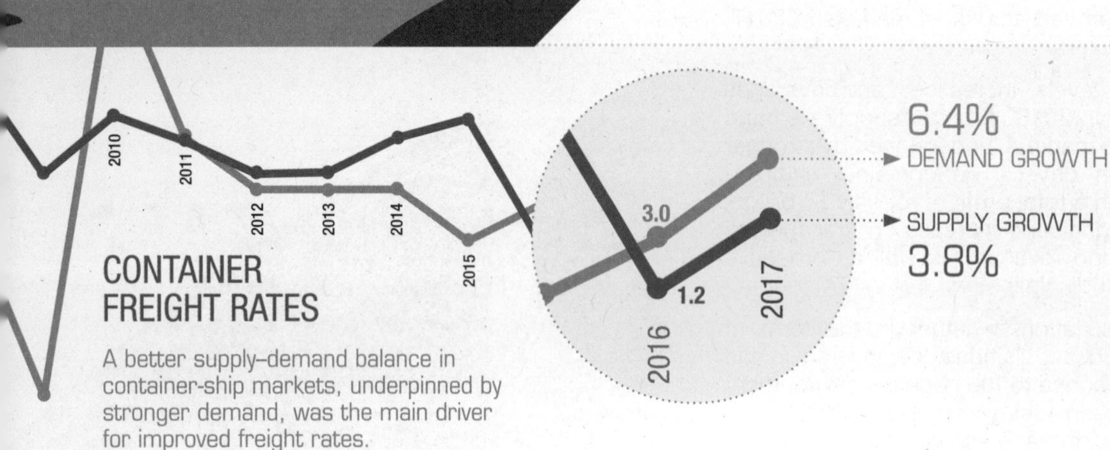


The container shipping industry ended 2017 with a total profit of

**$7 billion.**

Average earnings increased in all fleet segments, **$10,986** per day.

## DRY BULK FREIGHT RATES

**recorded a remarkable surge**, which translated into clear gains for carriers, thereby compensating the depressed earnings of 2016.

Seaborne dry bulk growth:
**4.4%**

Bulk carrier fleet growth:
**3%**

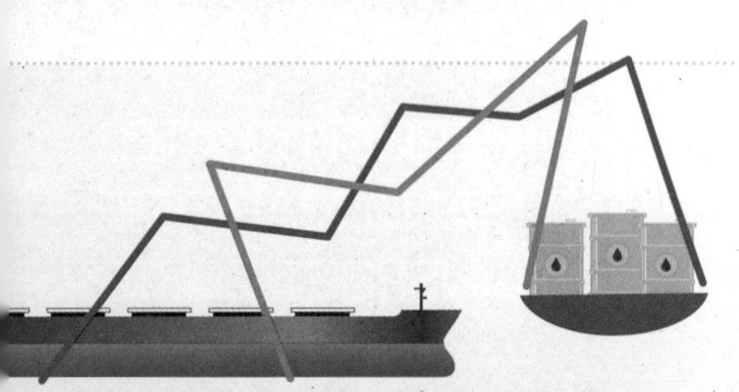

## TANKER FREIGHT RATES

remained under pressure, mainly due to an increase in vessel supply that grew at a faster rate than demand growth.

| Baltic Exchange dirty tanker index | Baltic Exchange clean tanker index |
|---|---|
| **8%** growth | **24%** growth |
| 787 points | 606 points |

## CONTAINER MARKET CONSOLIDATION

Consolidation, through **mergers and acquisitions** or **alliances**, persevered in the container industry in response to the negative environment and losses experienced by the industry in recent years.

### MERGERS AND ACQUISITIONS

Their share has increased further with the completion of the operational integration of the new mergers in 2018.

January 2018
Top 15 carriers

controlled
**70%**
of fleet capacity.

June 2018
Top 10 carriers

### ALLIANCES

Alliances reorganized to form three larger alliances of global carriers in 2017: **2M**, the Ocean Alliance and **"The" Alliance**, accounting for 93% of East–West lanes.

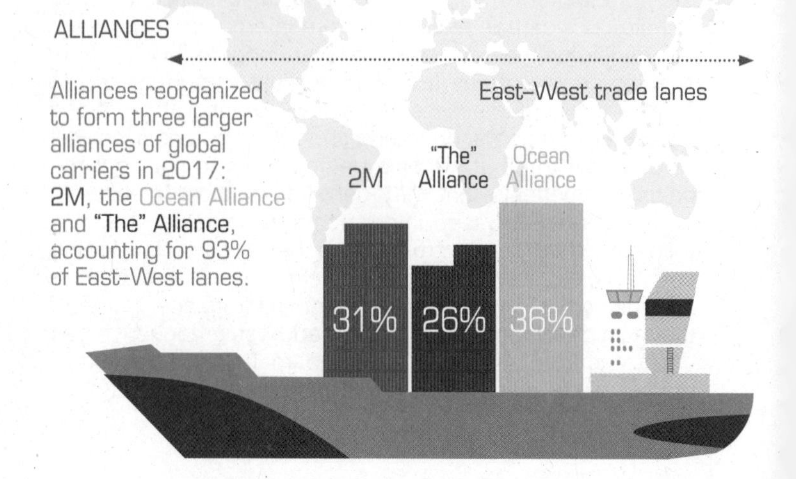

# A. CONTAINER FREIGHT RATES: CONSIDERABLE MARKET IMPROVEMENTS

## 1. Overview

The container freight market improved considerably, following a difficult market environment in 2016. As illustrated in figure 3.1, global container demand grew at 6.4 per cent in 2017, taking total volumes to an estimated 148 million TEUs. The strong development in global container shipping demand in 2017 reflects a fundamental improvement in the global economic environment. Demand growth was particularly high in the first three quarters of the year, although it slowed down in the last quarter. UNCTAD projects global containerized trade to expand at a compound annual growth rate of 6.4 per cent in 2018 supported by the positive economic trends (see chapter 1).

Global supply of container ship-carrying capacity, on the other hand, grew at an estimate of 2.8 per cent, bringing the total global capacity to 256 million dwt (see chapter 2). Although supply growth was relatively moderate, the container market continued, nevertheless, to struggle with the delivery of mega container ships and surplus capacity among the larger vessels (exceeding 14,000 TEUs). World fleet capacity is projected to rise by 3 per cent in 2018 (see chapter 2).

Eventhough the supply of global container ship capacity continued in 2017, freight rates made a remarkable recovery from the lows recorded in 2016. This performance was supported by the upturn in the global demand for container transport services in 2017 across all trade lanes. As shown in table 3.1, freight rates on the mainlane trades routes went up, although they remained volatile, with a drop in the second half due to low demand growth. The surge was driven mainly by positive market trends in the developed regions. During the year, the United States and the European Union recorded economic growth and higher import demand (see chapter 1). Average trans-Pacific spot freight rates increased by 16.7 per cent, with the Shanghai–United States West Coast routes averaging $1,485 per 40-foot equivalent unit (FEU). Rates on the Shanghai–United States East Coast route increased by 17.3 per cent over 2016 and averaged $2,457 per FEU. On the Shanghai–Northern Europe route, average rates stood at $876 per TEU, up by 27 per cent, whereas Shanghai–Mediterranean rates averaged $817 per TEU, an increase of 19.4 per cent over the previous year.

**Figure 3.1 Growth of demand and supply in container shipping, 2007–2017**
(Percentage)

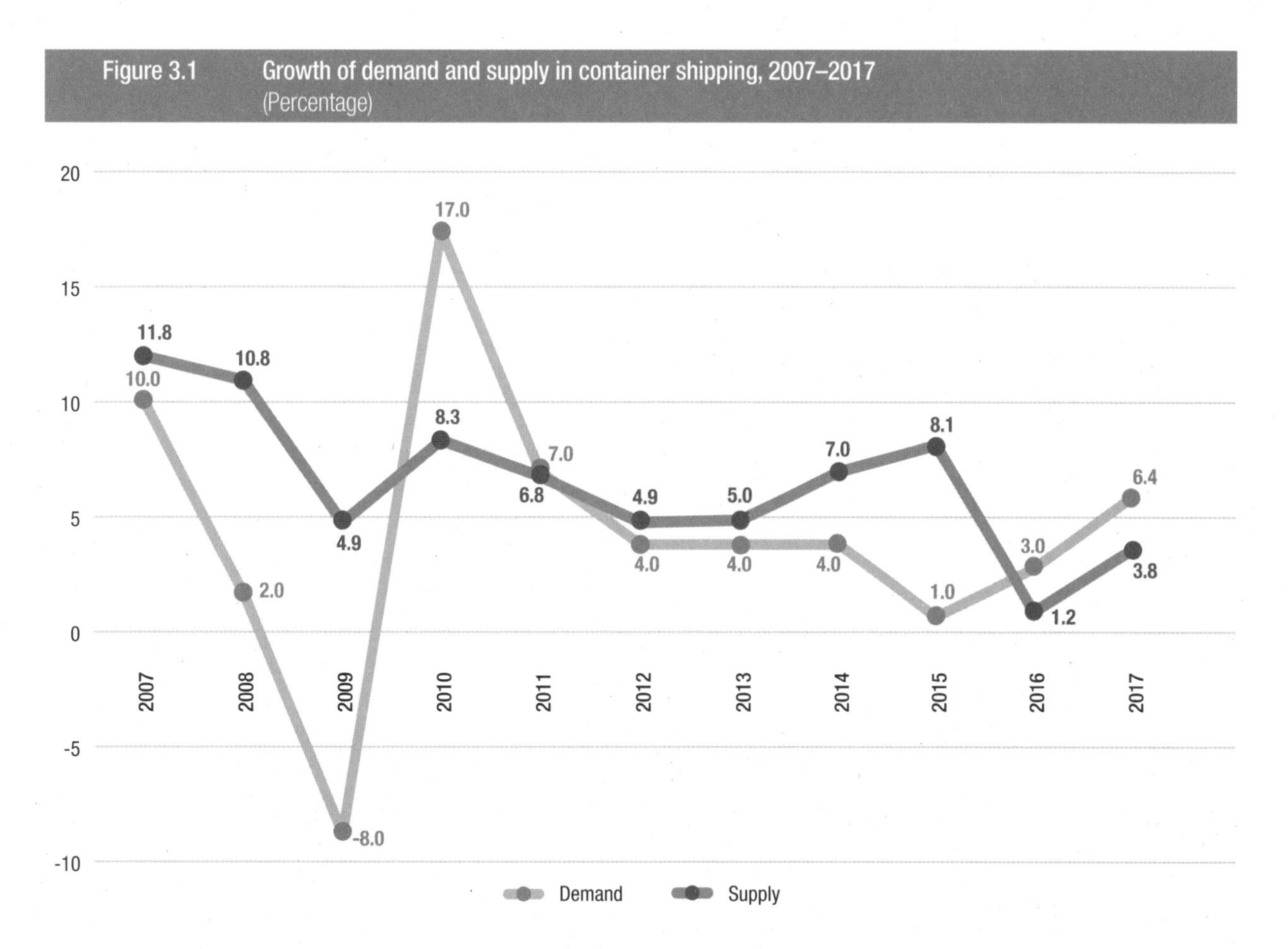


*Source:* UNCTAD secretariat calculations, based on data from chapter 1, figure 1.5 for demand and Clarksons Research, *Container Intelligence Monthly*, various issues, for supply.

*Notes:* Supply data refer to total capacity of the container-carrying fleet, including multipurpose vessels and other types of vessel with some container-carrying capacity. Demand growth is based on million TEU lifts.

On the non-mainlane routes, robust growth in all trade clusters supported the positive development of freight rates, which rose sharply in 2017, outperforming those on the mainlane trade routes. Among the North–South routes, the Shanghai–South Africa (Durban) freight rates averaged $1,155 per TEU, an increase of almost 98 per cent compared with 2016. The Shanghai–South America (Santos) annual freight rates reached an average of $2,679 per TEU, an increase of 62.7 per cent over the 2016 average. These surges were mainly driven by large growth in demand from oil and commodity-exporting countries following the 2017 improvements in the commodity price environment (see chapter 1).

With regard to the intra-Asian routes, the Shanghai–Singapore route averaged $148 per TEU, compared with $70 per TEU in 2016, a 111.4 per cent increase. These rates were supported by continued positive trends in the Chinese economy, as well as in other emerging economies in the region.

**Table 3.1 Container freight markets and rates, 2010–2017**

| Freight market | 2010 | 2011 | 2012 | 2013 | 2014 | 2015 | 2016 | 2017 |
|---|---|---|---|---|---|---|---|---|
| **Trans-Pacific** | **(Dollars per 40-foot equivalent unit)** | | | | | | | |
| Shanghai–United States West Coast | 2 308 | 1 667 | 2 287 | 2 033 | 1 970 | 1 506 | 1 272 | 1 485 |
| Percentage change | 68.2 | -27.8 | 37.2 | -11.1 | -3.1 | -23.6 | -15.5 | 16.7 |
| Shanghai– United States East Coast | 3 499 | 3 008 | 3 416 | 3 290 | 3 720 | 3 182 | 2 094 | 2 457 |
| Percentage change | 47.8 | -14.0 | 13.56 | -3.7 | 13.07 | -14.5 | -34.2 | 17.3 |
| **Far East–Europe** | **(Dollars per 20-foot equivalent unit)** | | | | | | | |
| Shanghai–Northern Europe | 1 789 | 881 | 1 353 | 1 084 | 1 161 | 629 | 690 | 876 |
| Percentage change | 28.2 | -50.8 | 53.6 | -19.9 | 7.10 | -45.8 | 9.7 | 27.0 |
| Shanghai–Mediterranean | 1 739 | 973 | 1 336 | 1 151 | 1 253 | 739 | 684 | 817 |
| Percentage change | 24.5 | -44.1 | 37.3 | -13.9 | 8.9 | -41.0 | -7.4 | 19.4 |
| **North–South** | **(Dollars per 20-foot equivalent unit)** | | | | | | | |
| Shanghai–South America (Santos) | 2 236 | 1 483 | 1 771 | 1 380 | 1 103 | 455 | 1 647 | 2 679 |
| Percentage change | -8.0 | -33.7 | 19.4 | -22.1 | -20.1 | -58.7 | 262.0 | 62.7 |
| Shanghai–Australia/ New Zealand (Melbourne) | 1 189 | 772 | 925 | 818 | 678 | 492 | 526 | 677 |
| Percentage change | -20.7 | -35.1 | 19.8 | -11.6 | -17.1 | -27.4 | 6.9 | 28.7 |
| Shanghai–West Africa (Lagos) | 2 305 | 1 908 | 2 092 | 1 927 | 1 838 | 1 449 | 1 181 | 1 770 |
| Percentage change | 2.6 | -17.2 | 9.64 | -7.9 | -4.6 | -21.2 | -18.5 | 49.9 |
| Shanghai–South Africa (Durban) | 1 481 | 991 | 1 047 | 805 | 760 | 693 | 584 | 1 155 |
| Percentage change | -0.96 | -33.1 | 5.7 | -23.1 | -5.6 | -8.8 | -15.7 | 97.8 |
| **Intra-Asian** | **(Dollars per 20-foot equivalent unit)** | | | | | | | |
| Shanghai–South-East Asia (Singapore) | 318 | 210 | 256 | 231 | 233 | 187 | 70 | 148 |
| Percentage change | | -34.0 | 21.8 | -9.7 | 0.9 | -19.7 | -62.6 | 111.4 |
| Shanghai–East Japan | 316 | 337 | 345 | 346 | 273 | 146 | 185 | 215 |
| Percentage change | | 6.7 | 2.4 | 0.3 | -21.1 | -46.5 | 26.7 | 16.2 |
| Shanghai–Republic of Korea | 193 | 198 | 183 | 197 | 187 | 160 | 104 | 141 |
| Percentage change | | 2.6 | -7.6 | 7.7 | -5.1 | -14.4 | -35.0 | 35.6 |
| Shanghai–Hong Kong SAR | 116 | 155 | 131 | 85 | 65 | 56 | 55 | — |
| Percentage change | | 33.6 | -15.5 | -35.1 | -23.5 | -13.8 | -1.8 | — |
| Shanghai–Persian Gulf/ Red Sea | 922 | 838 | 981 | 771 | 820 | 525 | 399 | 618 |
| Percentage change | | -9.1 | 17.1 | -21.4 | 6.4 | -36.0 | -24.0 | 54.9 |

*Source:* Clarksons Research, *Container Intelligence Monthly*, various issues.
*Note:* Data based on yearly averages.
*Abbreviation:* SAR, Special Administrative Region

**Figure 3.2 New ConTex index, 2010–2018**

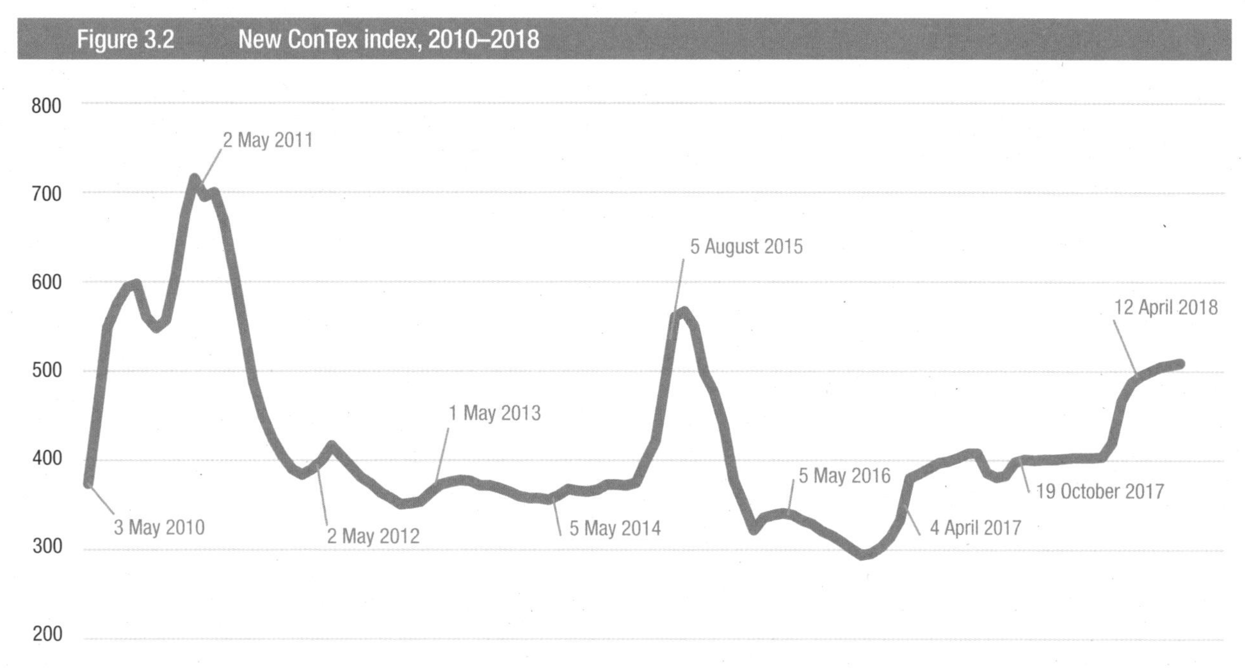


*Source:* UNCTAD secretariat, based on data from the New ConTex index of the Hamburg Shipbrokers Association.
*Notes:* The New ConTex is based on assessments of the current day charter rates of six selected container ship types, which are representative of their size categories: Types 1,100 TEUs and 1,700 TEUs with a charter period of one year, and Types 2,500, 2,700, 3,500 and 4,250 TEUs with a charter period of two years. Index base: October 2007 = 1,000 points.

In an effort to address overcapacity and absorb the impact of surplus capacity, slow steaming and cascading were strongly maintained by carriers in 2017. Slow steaming is estimated to have absorbed some 3 million TEUs of nominal capacity since the end of 2008 (Clarksons Research, 2018a). Cascading capacity resulted in increasing the redeployment of larger ships across trade lanes (Clarksons Research, 2018a). Larger ships are deployed on mainlane trade routes, which require carriers to balance capacity and distribute ships onto secondary lanes, such as the North–South trade lanes. At the same time, and as noted in chapter 2, scrapping of vessels remained significantly high – 4.5 million gross tons were demolished in 2017. The average age of scrapped vessels stood at 21 years in 2017 (Clarksons Research, 2018a), an average that has been steadily falling over the years, from 33 years in 2008 to 26 years in 2016 (Hellenic Shipping News, 2017). Supported by demand growth, the level of container ship idling, which represented about 7 per cent of the active fleet in late 2016 and early 2017, reached about 2 per cent in late 2017 (Barry Rogliano Salles, 2018).

In line with developments concerning demand, supply and spot rates, the shipping charter market also improved in 2017, as rates increased in most sectors over the year, with some volatility and variation across vessel sizes. The 12-month charter rate increased to an average of 378 points, compared with 325 average points in 2016 (figure 3.2). Partly sustained by stronger container demand, this surge reflected the start of the new alliance structures requiring carriers to charter vessels to fill gaps while their networks were being formed. Another factor that drove up the rates was that carriers needed to fill short-term capacity requirements, while awaiting the delivery of new ships. Orient Overseas Container Line, for instance, hired some ships with a capacity of 11,000 TEUs from Costamare to operate on the Asia–North Europe trade route pending the arrival of new units with a capacity of 20,000 TEUs (JOC.com, 2017).

The container ship charter market got off to a good start in 2018. The new ConTex index increased to an average of close to 500 points in April 2018, the highest since August 2015. Nevertheless, there are still concerns about the potential cascading effect of larger vessel sizes with the delivery of new mega vessels, as well as the impact of market consolidation on vessel employment by major carriers, which may seek to rationalize supply capacity, or use their own tonnage and seek to off-hire chartered ships to control fleet supply (The Loadstar, 2018).

## 2. Global container shipping: A year of positive growth and profits

Following a year of losses in 2016, the container shipping industry ended 2017 with a total profit of roughly $7 billion (Drewry, 2018), driven mainly by a significant increase in transported volumes, freight rates and revenue, as well as proactive and disciplined operational management. CMA CGM recorded the best operating results in container shipping, with core earnings before interest and taxes reaching $1.575 billion (CMA CGM, 2018a; CMA CGM, 2018b), followed by Maersk with gains of $700 million (A. P. Moller–Maersk, 2018). Hapag-Lloyd ranked third, with €410.9 million (about $480 million) (Hapag-Lloyd, 2018). The financial performance and relevant activities of selected carriers is summarized in box 3.1.

**Box 3.1 Financial performance and relevant activities of the top three shipping lines, 2017**

**CMA CGM**

In 2017, the financial situation of CMA CGM was characterized by an increase in revenue of 32.1 per cent, reaching $21.1 billion. Due to an increase in freight rates and volumes, its average revenue per TEU rose by 9 per cent over that of 2016.

Its core earnings before interest and taxes amounted to $1.575 billion, with a margin of 7.5 per cent core earnings before interest and taxes, up 7.3 points from the previous year. This was made possible by a rise in average revenue per TEU transported and control of unit costs, which rose slightly by 1.6 per cent, despite a sharp rise in fuel prices.

The shipping line recorded a net profit of $701 million in 2017, a sharp increase compared with 2016, when it posted a loss of $452 million.

CMA CGM carried nearly 19 million containers, an increase of 21.1 percent over 2016. This increase is driven by contributions of all the shipping lines operated by the Group, in addition to the first full-year contribution of American President Lines, which carried more than 5 million TEUs and contributed $340 million to the Group's operating income.

In October 2017, CMA CGM acquired Sofrana, an operator in the South Pacific islands, and in December, closed the acquisition of Mercosul Line, one of the main players in Brazil's domestic container shipping market.

On 1 April, the Ocean Alliance, the world's largest operational shipping alliance, boasting 40 services and more than 320 ships, was launched.

In 2017, the Group accelerated its digital transformation. Numerous initiatives have already been launched as part of the establishment of CMA CGM Ventures, which is devoted to corporate investments in innovative technologies, the development of partnerships with major e-commerce groups and other similar activities.

In 2017, CMA CGM took delivery of the *Antoine de Saint-Exupery*, the largest container ship flying the French flag. The vessel has a number of new environmentally friendly features, including an IMO-required ballast water treatment system to mitigate the transport of marine-invasive species. The vessel benefits from premium technologies such as the Becker Twisted Fin allowing improvements in propeller performance, helping reduce significantly the energy expenditure for a 4 per cent reduction in carbon dioxide emissions and a new-generation engine that significantly reduces oil consumption (less 25 per cent) and fuel consumption for a 3 per cent average reduction of carbon dioxide emissions.[a]

**Maersk**

Maersk's revenue increased by 14.9 per cent in 2017 to reach $23.8 billion, compared with $20.7 billion in 2016. This was mainly attributed to an increase in volumes and an average freight rate of 11.7 per cent.

Earnings before interest and taxes stood at $700 million in 2017, compared with a $396 million loss in 2016. Maersk reported a return to profit of $521 million in 2017, as opposed to a loss of $384 million in 2016. These results benefited from the shipping company's higher revenue and a unit cost at fixed bunker price almost on a par with results in 2016. The unit cost at fixed bunker price was, however, negatively affected by a cyberattack in the third quarter of 2017, as well as decreased headhaul utilization and lower backhaul volumes. Total unit costs increased by 4.9 per cent in 2017, attributed in large part to an increase in the average price of bunker fuel.

Transported volumes grew from 10.41 million FEUs in 2016 to 10.73 million FEUs in 2017, an increase of 3.0 per cent, despite the negative impact of the cyberattack. The increase in volume was driven by an increase in East–West volumes of 2.4 per cent; North–South volumes, of 2.2 per cent; and intraregional volumes, of 7.3 per cent.

The acquisition of Hamburg Süd and the divestment of Mercosul Line were completed in December 2017.

In the area of digitalization, Maersk launched a remote container management programme for customers in July 2017, which provides the location of refrigerated containers throughout its journey, as well as the atmospheric conditions inside each container. In January 2018, the A. P. Moller–Maersk Group and International Business Machines (IBM) announced their intent to establish a joint venture to provide more efficient solutions to digitalize supply chain documentation and secure methods for conducting global trade using blockchain technology.

Maersk took delivery of 5 of 11 second-generation Triple-Es and 4 of 9 vessels with a capacity of 15,200 TEUs, which had been ordered in 2015. The new vessels replaced older and less efficient vessels, and as part of this process, Maersk recycled 16 vessels in 2017.

**Hapag-Lloyd**

On 24 May 2017, the merger of Hapag-Lloyd and the United Arab Shipping Company took place, and operational integration of the United Arab Shipping Company Group was completed in late November. Owing to an increase in transport volumes and in average freight rates, as well as to the inclusion of the United Arab Shipping Company Group, Hapag-Lloyd reported €9.97 billion in revenue, compared with €7.73 billion in 2016. Freight rates averaged $1,051 per TEU, exceeding the previous year's level by 1.4 per cent (2016: $1,036 per TEU). Freight rate increases, particularly in the Far East, Middle East and Latin America trade routes, had a positive impact on earnings.

Hapag-Lloyd's operating results (earnings before interest and taxes) stood at €410.9 million (about $480 million) clearly above the previous year's level of €126.4 million. This resulted in an earnings before interest and tax margin of 4.1 per cent (prior year: 1.6 per cent).

Transported volumes rose by 29 per cent in 2017, reaching 9.803 million TEUs, compared with 7.599 million TEUs in 2016, primarily as a result of the acquisition of the United Arab Shipping Company. This also led to a significant increase in the average ship size and a reduction in the average age of vessels.

Transport expenses rose by €1,626 million to €7,990 million, compared with €6,364 million in 2016. This represents an increase of 25.5 per cent that is primarily due to the acquisition of the United Arab Shipping Company Group and related growth in transport volumes and higher bunker prices. At 19.9 per cent, transport expenses, not including bunker costs, increased at a much lower rate than the increase in transport volumes (29.0 per cent).

Container shipping utilizes information technology in processes such as yield management, shipping quotations, cargo volume management, the design of new shipment services and operation of empty legs. A digital channel and incubation unit was established in 2017 to develop new, digitally available services and business models.

*Source:* Carriers' annual reports (2017) and websites.

[a] https://shipinsight.com/articles/cma-cgm-takes-delivery-20600-teu-flagship-antoine-de-saint-exupery.

## 3. Consolidation persevered in the container market

In 2017, consolidation, through mergers and acquisitions or alliances persevered in the container industry in response to the negative environment and losses experienced by the industry in recent years. The world's leading container shipping lines recorded an estimated collective operating loss of $3.5 billion in 2016, their first annual deficit since 2011 (Lloyd's Loading List, 2017).

Key mergers and acquisitions in 2018 involved the merger of the Japanese container ship operator groups "K" Line (Kawasaki Kisen Kaisha), Mitsui Osaka Shosen Kaisha Lines and NYK Lines (Nippon Yusen Kabushiki Kaisha) to form Ocean Network Express and the planned merger of Orient Overseas Container Line with the China Ocean Shipping Company. Ocean Network Express will rank sixth in terms of global ranking by vessel capacity — a combined 1.53 million TEUs (above Evergreen's 1.1 million TEUs and just behind Hapag-Lloyd's 1.55 million TEUs) (see chapter 2). As of January 2018, the top 15 carriers accounted for 70.3 per cent of all capacity. The five leading carriers – Maersk, Mediterranean Shipping Company, CMA CGM, China Ocean Shipping Company and Hapag-Lloyd – control more than 50 per cent of market capacity. Their share has increased further with the completion of the operational integration of the new mergers in 2018, as the top 10 shipping lines controlled almost 70 per cent of fleet capacity as of June 2018 (see chapter 2).

Mergers, if well-conceived and accompanied by effective executional strategies, can deliver greater value and help carriers improve performance and operational synergies. For instance, cost synergies from the merger of Hamburg Süd and Maersk are expected to range from $350 million to $400 million by 2019, primarily from integrating and optimizing the networks, as well as standardizing procurement procedures (A. P. Moller–Maersk, 2018). Hapag-Lloyd, which merged with the United Arab Shipping Company in May 2017, estimates that it will generate $435 million in cost synergies from 2019 as a result of the merger (Hapag-Lloyd, 2017). China Ocean Shipping Company and Orient Overseas Container Line also foresee significant cost synergies, while maintaining separate brands (see www.hellenicshippingnews.com/container-shipping-more-mergers-better-mergers/).

Alliances of global carriers were restructured in 2017 to form three larger ones: 2M, the Ocean Alliance and "The" Alliance.[1] This reshuffling resulted in a highly concentrated market structure, mainly in the main trade lanes, where the three alliances collectively account for around 93 per cent of the East–West routes, leaving 7 per cent for the other smaller global and regional carriers (The Maritime Post, 2018). With regard to the deployed capacity of alliances on the three major East–West lanes combined, figure 3.3 shows that the Ocean Alliance is the largest, with a 36 per cent share of the market, followed by 2M, with 31 per cent, and "The" Alliance, with 26 per cent, based on data as at May 2018. The remaining 7 per cent is held by non-alliance members, whose deployed capacity varies by routes operated.

Compared with 2014, the average number of services provided by all liner shipping operators fell by 6 per cent to reach 474 in the second quarter of 2018, from 504 in the first quarter of 2014 (The Maritime Post, 2018). The number of services provided by members of the alliances, however, increased from 150 in in the first quarter of 2014 to 297 in the second quarter of 2018 (98 per cent increase). In contrast, services offered by other operators not members of an alliance decreased by 46.2 per cent, from 431 in the first quarter of 2014 to 232 services in the second quarter of 2018 (The Maritime Post, 2018). Although it is not clear whether the decrease in services has negatively affected the options available to shippers, this is a potentially

Figure 3.3 Capacity deployed by alliances in principal East–West trade lanes, 2018
(Percentage)

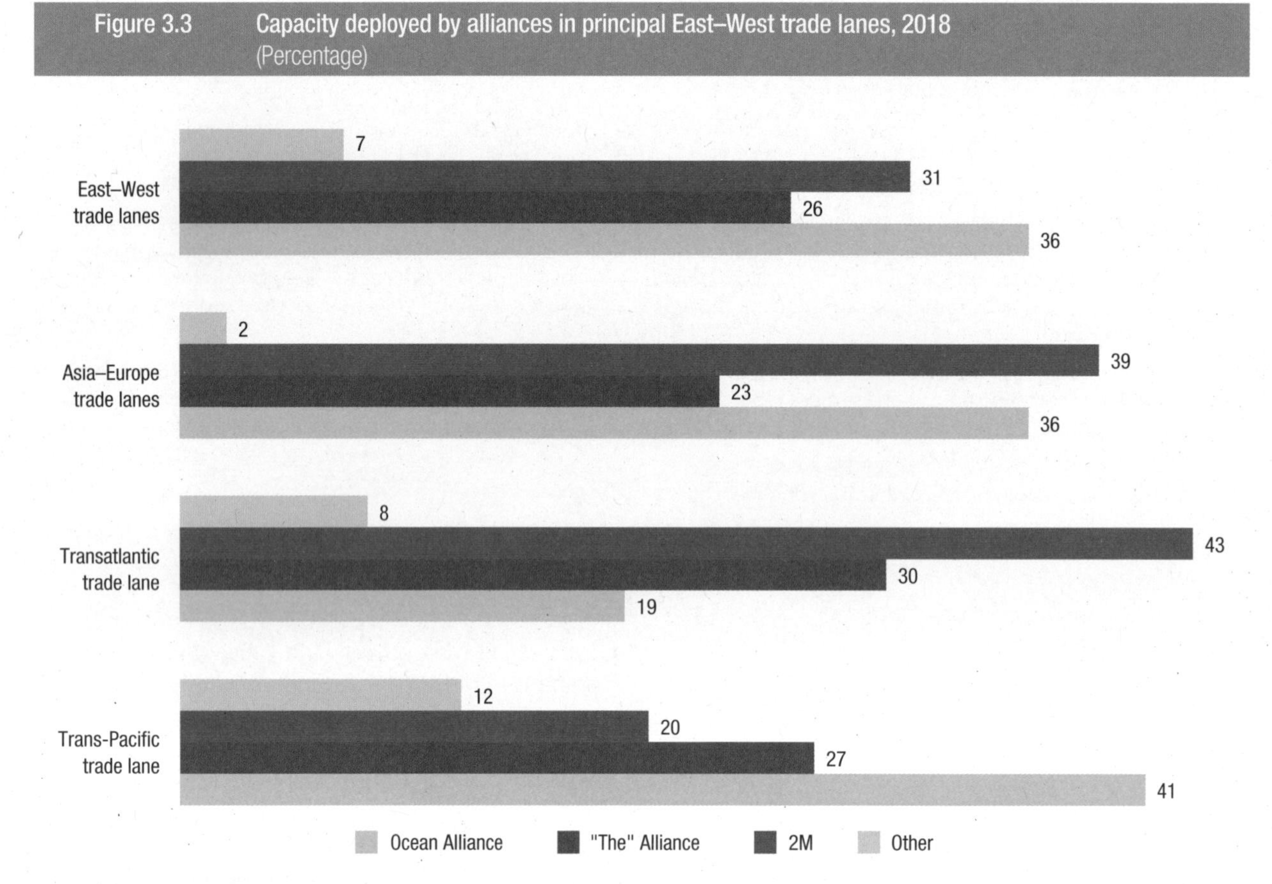


*Source:* MDS Transmodal, 2018.
*Note:* Data as of May 2018.

worrisome trend if sustained. The impact of increasing consolidation is also felt by smaller operators that do not belong to an alliance. Their share in deployed capacity is 2 per cent in the Asia–Europe trade lanes, 8 per cent in the transatlantic trade lane and 12 per cent in the trans-Pacific trade lane (figure 3.3). However, in many cases, many of these operators have a more regional focus and tend to be more active in niche markets or individual routes.

For shippers, increased consolidation means fewer carrier choices, less competition and ultimately, carriers in a better position to influence market prices and increase freight rates (see chapter 1). However, there has been no evidence of this having been achieved in 2017, as alliances' operations are still being defined, and the industry is still struggling to achieve economies of scale and lower operational costs, while improving supply-capacity utilization on certain routes that jeopardize the balance of market fundamentals in an uncertain world. Yet, and as noted in the two previous editions of the *Review of Maritime Transport*, there is still a risk that growing concentration and consolidation of the market will distort competition and will be detrimental to the market, freight rates and shippers. Therefore, the oversight role of competition authorities and regulators should be strengthened and their capacities reinforced to monitor the evolution of current alliances and to review mergers and acquisitions so as to ensure fair competition and prevent anticompetitive practices. Such practices may create a significant impact on smaller players with weak bargaining power, notably those from developing countries. At the same time, authorities and shippers would need to consider the quality, reliability and variety of services provided to shippers in addition to the effects of price competition. Competition authorities should also consider the effects on factors such as the range and quality of services, frequency of ships, range of ports serviced and reliability of schedules (UNCTAD, 2018).

## B. DRY BULK FREIGHT RATES: NOTABLE RECOVERY

The dry bulk market underwent a remarkable recovery in 2017 . Growth in demand for seaborne dry bulk surpassed the fleet growth, as demand for commodities went up, while the surplus of vessels gradually continued to diminish. As noted in chapter 1, seaborne dry cargo shipments increased by 4.4 per cent in 2017, up from 2.0 per cent in 2016. Bulk carrier fleet growth, on the other hand, remained manageable at 3.0 per cent in 2017; deliveries declined to almost 20 million gross tons, and scrapping activities increased to more than 8 million gross tons (see chapter 2).

**Figure 3.4 Baltic Exchange Dry Index, 2003–2018**

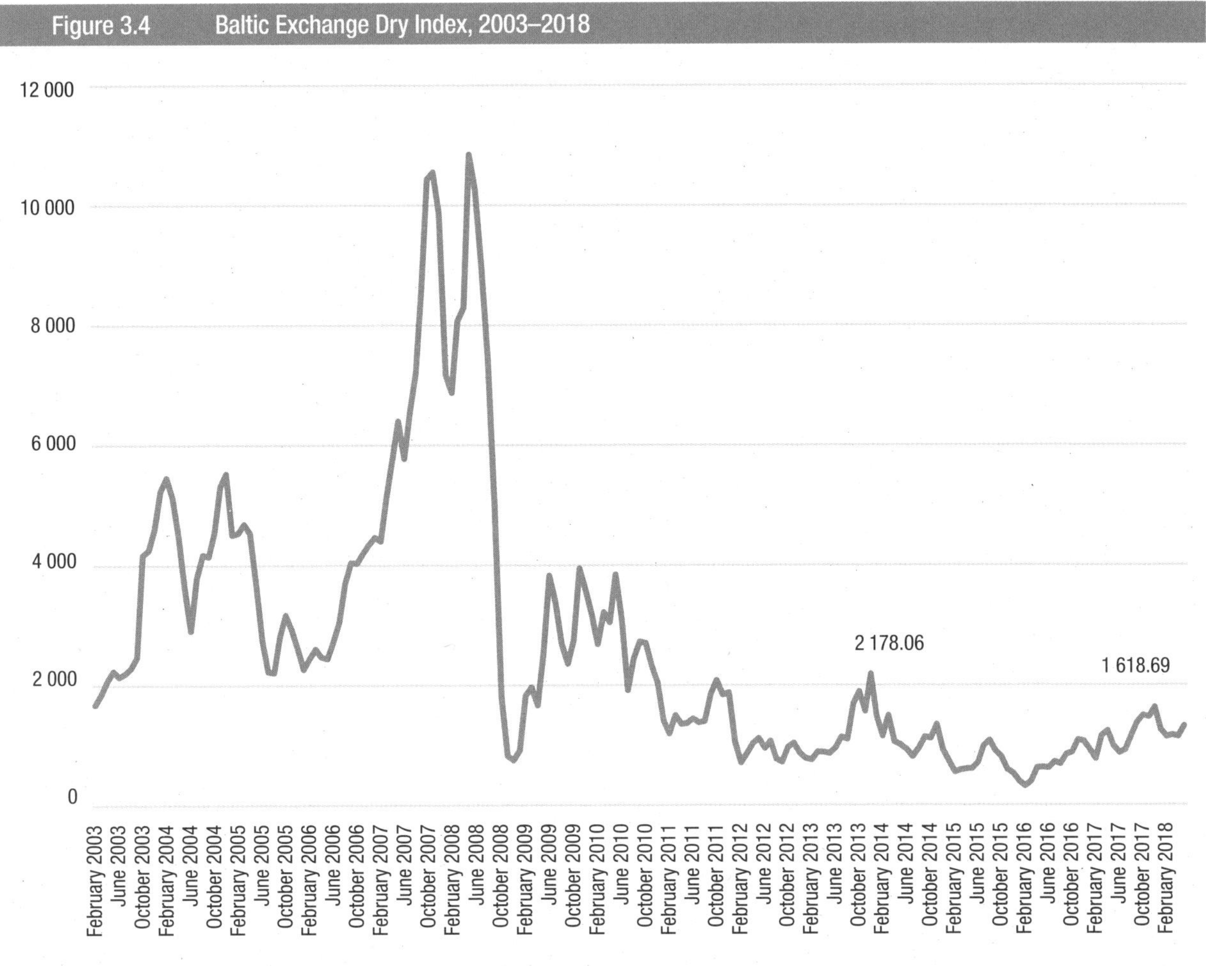


*Source:* UNCTAD secretariat calculations, based on data from the Baltic Exchange.
*Notes:* The Index is made up of 20 key dry bulk routes measured on a time charter basis and covers Handysize, Supramax, Panamax and Capesize dry bulk carriers, which carry commodities such as coal, iron ore and grain. Index base: 1 November 1999 = 1,334 points.

Consequently, the Baltic Exchange Dry Index rebounded, especially after having experienced one of the weakest years in 2016 since the financial crisis. As shown in figure 3.4, the Index averaged about 1,153 points, reaching a peak of 1,619 points in December 2017, the highest level since 2013, when it had reached 2,178 points.

As a result, average earnings increased in all fleet segments, averaging $10,986 per day in 2017, up by 77 per cent from the depressed levels of 2016 (Clarksons Research, 2018b). The sector experienced a strong rebound in charter rates as growth in demand for commodities exceeded fleet expansion.

## 1. Capesize

The Capesize market improved significantly in 2017, driven largely by the surge in growth in the iron ore imports of China and a rebound in coal trade, which helped curb the level of supply capacity. Charter and freight rates improved substantially, as illustrated by the average Baltic Capesize Index of the four and five time charter routes, which recorded a high daily level of $14,227 and $15,291, respectively, twice the average rates of 2016 (figure 3.5).

## 2. Panamax

Market conditions in the Panamax sector also improved markedly from the historically depressed levels of 2016, supported by an improvement in the supply–demand balance. The Baltic Panamax Index of the four time charter routes averaged at $10,570 per day in 2017, up by 75 per cent from the 2016 average. Improved demand supported by an expansion in coal and grain shipments and firm growth in key minor bulk commodities trade, prompted positive trends. At the same time, growth on the supply side remained moderate as the fleet increased by 2.7 per cent (Clarksons Research, 2018b).

## 3. Handysize and Supramax

Similarly, Handysize market conditions improved in 2017. The Baltic Supramax Index of the six time charter routes averaged $9,185 per day, up by 46 per cent ($6,270 per day), and the Baltic Handysize Index of the six time charter routes averaged $7,662 per day from $4,974 per day in 2016, a 54 per cent increase over 2016 (figure 3.5). More positive demand-side trends (growth in coal,

**Figure 3.5 Daily earnings of bulk carriers, 2009– 2018**
(Dollars per day)

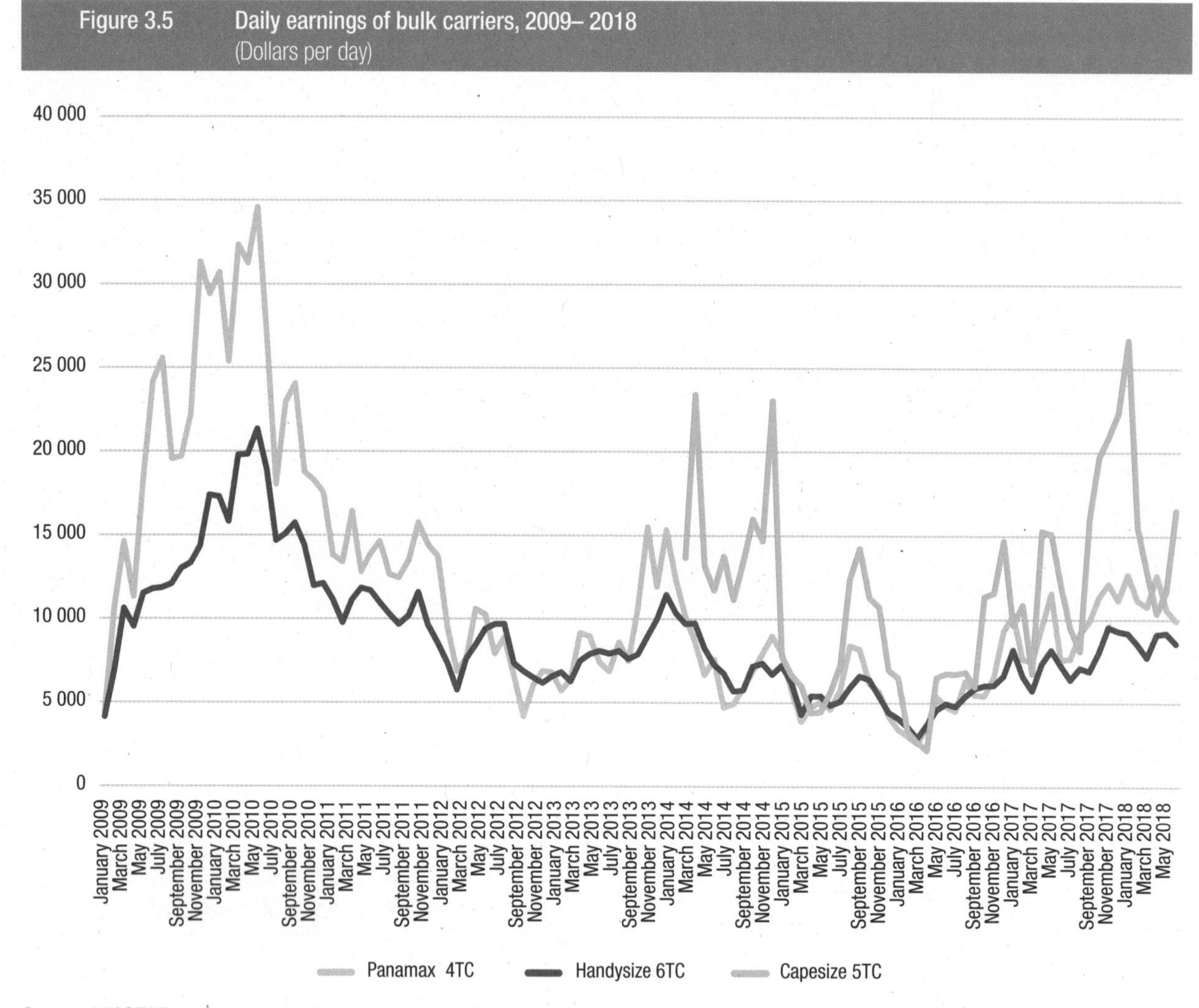


*Source:* UNCTAD secretariat calculations, based on data from Clarksons Research Shipping and the Baltic Exchange.
*Abbreviations:* Panamax 4TC, average rates of the four time charter routes; Capesize 5TC, average rates of the five time charter routes; Handysize 6TC, average rates of the six time charter routes.

grain and minor bulk trade) and continued limited supply growth helped support these improvements. In 2018, improvements to the fundamental balance will sustain positive growth for dry bulk shipping rates.

## C. TANKER FREIGHT RATES: A CHALLENGING YEAR

Overall, 2017 proved to be a challenging year for the tanker market, mainly because of the pressure faced by markets from continuous growth in supply capacity, particularly in the crude tanker sector that was matched by a relative deceleration in demand growth. It is estimated that global tanker trade expanded at an annual average growth rate of 3.0 per cent in 2017 (see chapter 1); the crude oil tanker fleet grew by 5 per cent and the product tanker fleet grew by 4.2 per cent (Clarksons Research, 2018c). Rapid growth in the capacity of tankers carrying crude oil and products has further affected market balance, particularly in the crude oil sector.

As a result, the Baltic index for crude oil (Baltic Exchange dirty tanker index) recorded 8 per cent growth in 2017, reaching 787 points. The Baltic Exchange clean tanker index progressed by 24 per cent from the low level of 2016, reaching 606 points (table 3.2).

Freight rates also remained weak for both crude and products transports during most parts of 2017.

Earnings in the tanker sector weakened further over 2017 (figure 3.6), particularly in the crude tanker sector. Average spot earnings in all sectors fell significantly, reaching an average of $11,655 per day, a drop of 35 per cent from 2016 and the lowest annual average level in 20 years (Clarksons Research, 2018c). Performance on key crude tanker trades was poor, largely attributable to a reduction in Western Asia's exports in line with production cuts led by the Organization of the Petroleum Exporting Countries, coupled with rapid growth and oversupply in the crude tanker fleet (Hellenic Shipping News, 2018). For very large crude carriers, this was translated into low earnings averaging $17,800 per day, down by 57 per cent from 2016.

**Table 3.2 Baltic Exchange tanker indices, 2007–2018**

| | 2007 | 2008 | 2009 | 2010 | 2011 | 2012 | 2013 | 2014 | 2015 | 2016 | 2017 | Percentage change (2017/2016) | 2018 (first half year) |
|---|---|---|---|---|---|---|---|---|---|---|---|---|---|
| Dirty tanker index | 1124 | 1510 | 581 | 896 | 782 | 719 | 642 | 777 | 821 | 726 | 787 | 8 | 667 |
| Clean tanker index | 974 | 1155 | 485 | 732 | 720 | 641 | 605 | 601 | 638 | 487 | 606 | 24 | 577 |

*Source:* Clarksons Research, 2018d.

*Notes:* The Baltic Exchange dirty tanker index is an index of charter rates for crude oil tankers on selected routes published by the Baltic Exchange. The Baltic Exchange clean tanker index is an index of charter rates for product tankers on selected routes published by the Baltic Exchange. Dirty tankers generally carry heavier oils – heavy fuel oils or crude oil – than clean tankers. The latter generally carry refined petroleum products such as gasoline, kerosene or jet fuels, or chemicals.

In the product tanker sector, market conditions remained fairly steady at relatively weak levels. Supply continued to grow at a rate of 4.2 per cent in 2017. Meanwhile, volumes of refined petroleum products and gas increased by 3.9 per cent, supported by firm intra-Asian products trade and robust growth in Latin American imports (chapter 1). The cumulative effect of supply growth in recent years continued to depress earnings. Product tanker rates, which dropped sharply in 2016, remained at low but stable levels throughout 2017. A one-year time charter on a medium-range 2 tanker fluctuated between $12,500 and $14,500 per day.

As a result of poor market conditions, scrapping increased in the tanker sector and contributed about 11.2 million dwt in 2017, which is four times higher than 2016, when only about 2.5 million dwt were demolished (Clarksons Research, 2018c). This high level of demolition also continued into 2018.

In 2018, tanker trade volumes are projected to increase, although at a slightly slower pace than other market segments. However, oversupply capacity should be effectively managed to improve market balance and freight rates.

**Figure 3.6 Clean and dirty earnings, 2016–2018**

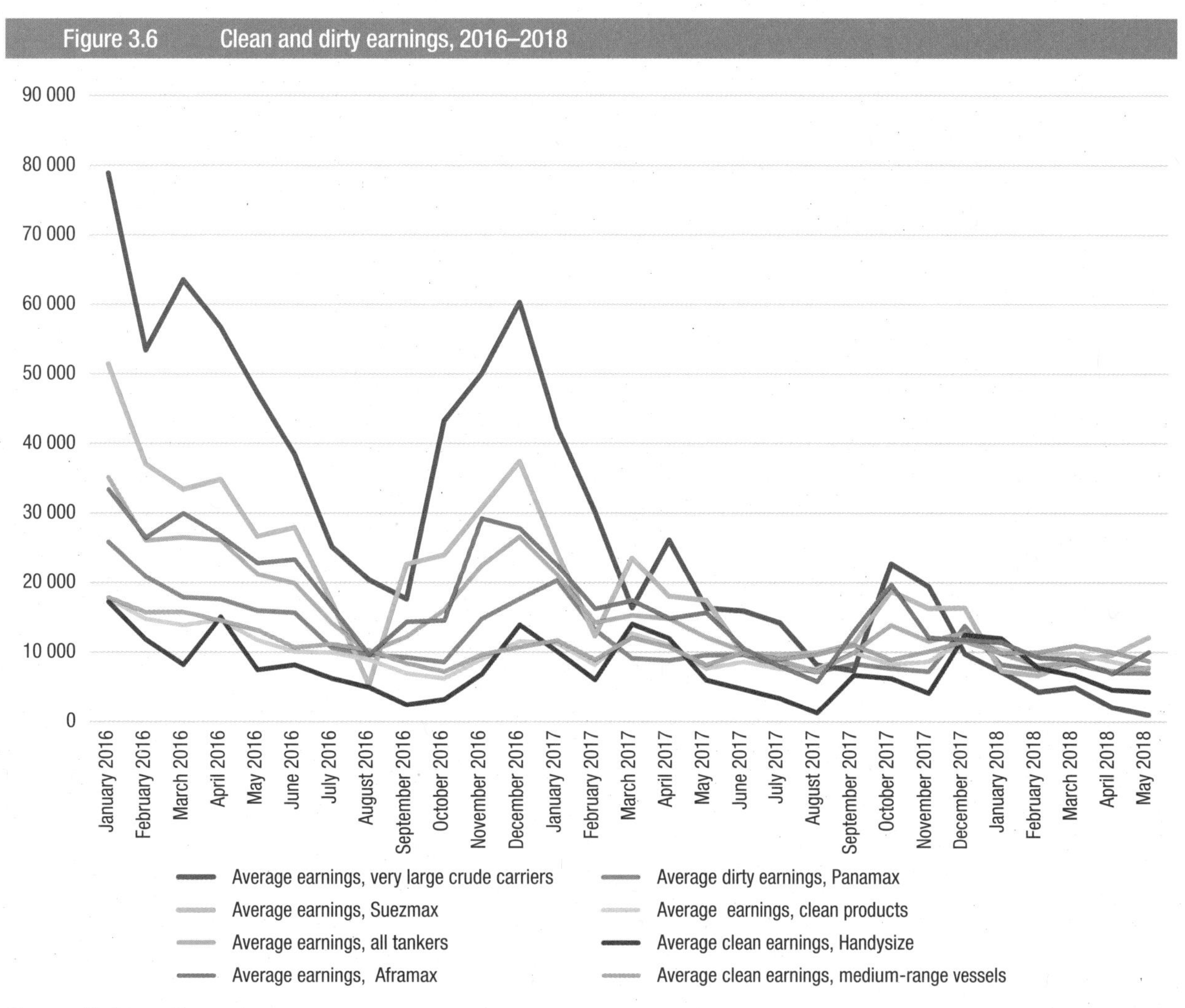


*Source:* Clarksons Research.

*Note:* Aframax, Suezmax and very large crude carriers were built circa 2000.

Reflecting positive trends in demand and better management of the supply side, global shipping freight rates improved, despite some variations by market segment. The overall outlook remains positive in view of improved market fundamentals. However, for these prospects to materialize, the prevailing downside risks need to be effectively contained.

Another key development to observe, from the perspective of carriers and shippers and their financial stance, is the current debate at IMO regarding the introduction of a set of short- to long-term measures to help curb carbon emissions from international shipping. The outcome of relevant negotiations and the specific design of any future instruments to be adopted may have implications for carriers, shippers, operating and transport costs, and costs for trade. It will therefore be important to assess those implications and consider the gains and benefits that may derive from future instruments, including market-based instruments in shipping. Further, it will be important to ascertain how they could be directed to address the needs of developing countries, especially in terms of their transport cost burden and their ability to access the global marketplace. In this context, the following section outlines some key measures taken at IMO to address greenhouse gas emissions from ships, as well as issues for consideration, particularly with regard to market-based instruments.

## D. GREENHOUSE GAS EMISSIONS REDUCTION IN SHIPPING: MARKET-BASED MEASURES

In April 2018, at the seventy-second session of the Marine Environment Protection Committee, IMO adopted a strategy on the reduction of greenhouse gas emissions from ships in line with the Paris Agreement under the United Nations Framework Convention on Climate Change and its ambition to maintain the global temperature rise well below 2 degrees Celsius above pre-industrial levels and to pursue efforts to limit the temperature increase even further to 1.5 degrees Celsius (see chapter 5). The IMO strategy sets out a vision to decarbonize the shipping sector and phase out greenhouse gas emissions from international shipping as soon as possible in this century, with the aim to reduce total annual greenhouse gas emissions by at least 50 per cent by 2050 compared with 2008 levels, while, at the same time, pursuing efforts towards phasing them out entirely. The strategy also sets to decrease the sector's average carbon intensity by at least 40 per cent until 2030, and 70 per cent by 2050.

Several short-, mid- and long-term measures are being considered as part of a comprehensive package of actions, including measures to improve energy efficiency and to stimulate the uptake of alternative fuels, while ensuring equity through the guiding principle of common but differentiated responsibilities and respective capabilities.[2] Market-based measures such as fuel levies and emissions trading systems are also considered part of the medium-term solutions (box 3.2).[3] Any set of measures that would be adopted by IMO would entail some financial implications for the sector. Consequently, the net impact of these multiple measures is likely to have some influence on transport rates and costs but how exactly this net impact will appear would require further analysis. This section will discuss some of the general concepts of market-based measures and its implication in the shipping sector. (For an assessment of some of the market-based measures proposals submitted to IMO between 2010 and 2012, see Psaraftis (2012).)

### 1. Policy levers for successful market-based measures

Similar to other measures, emissions-trading schemes and carbon levies have their advantages and disadvantages. It has yet to be determined at IMO

**Box 3.2 Market-based measures**

The market-based measures most commonly referred to are emissions-trading systems and carbon levies.

There are two main types of **emissions trading systems**:

- The cap-and-trade system, where a maximum amount of allowed emissions is determined (emissions cap), and emissions allowances (normally each one representing the right to emit one ton of carbon dioxide) are auctioned (market-based price setting-approach) or distributed for free according to specific criteria ("grandfathered").
- The baseline-and-credit system, where no maximum amount of emissions is set. An emissions intensity for emitting activities is set against a baseline, which can be business as usual or some proportion thereof. Polluters emitting less than the baseline would earn credits that they can sell to others who need them to comply with emission requirements.

A **carbon levy** directly fixes a price for carbon dioxide (usually per ton as in an emissions trading system) and can be applied as a fuel levy on the carbon content of fossil fuels. As opposed to an emissions trading system, the emissions reduction outcome is not predetermined but the carbon price is (non-market-based price setting).

*Sources:* Carbon Pricing Leadership Coalition, 2018; Organization for Economic Cooperation and Development, 2018.

whether, in addition to other policies (for example policies focused on efficiency or fuels), market-based measures are a cost-effective enabler of shipping decarbonization. Further, it is not clear what specification of market-based measures would be best suited to achieve the decarbonization target, while being politically acceptable to relevant stakeholders. The upsides and downsides of key policy levers of market-based measures are discussed in the following paragraphs, and an overview is provided in figure 3.7.

### *Price-setting mechanism*

Market-based price setting under an emissions cap has the implicit advantage of a guaranteed environmental outcome – only a predetermined amount of emission allowances are released into the market. The allowance price is then developed as a function of market demand (cap and trade) and fluctuates over time. With the price of emissions being directly set by the market, it adjusts automatically to the current costs of avoiding greenhouse gas emissions. A downside is the uncertainty of the price compared with a levy system. Existing emissions-trading schemes have a history of weak prices due to an oversupply of emissions certificates – too many allowances were allocated free of charge out of competitiveness concerns, and demand was overestimated, given unforeseen market developments such as the financial crisis of 2007 and an unexpectedly quick adoption of low-carbon technologies. Provisions to adjust the price were not part of the scheme architecture. As a result, the price signal was not as strong as expected to provide the desired incentive to invest in low-carbon technologies. In a high-demand scenario, on the other hand, prices may surge, especially when the sector comes close to reaching the emissions cap. Among the shortcomings of an emissions-trading scheme is the relative complexity of the system that could undermine smaller companies' competitiveness. For carbon levies, advantages and disadvantages are inverted: Investment security is higher, and transaction costs are lower, but the environmental outcome is not guaranteed. However, the choice between a fixed-quantity approach (emissions-trading system) and a fixed-price approach (levy) is not absolute. In emissions trading, the outcome is certain but the price will not be known in advance. With a fixed levy, the price is known but the effect on emissions is not. An emissions-trading system could have a floor price, and a levy could be regularly reset to reflect recent market developments.

### *Revenue generation*

In addition to the price level, the amount of revenues generated by market-based measures depends on whether emissions charges are calculated based on total or partial emissions. One approach is to require carriers to pay for all greenhouse gas emissions generated by bunker fuel combustion. Alternatively, only the difference to an emissions benchmark per ship

**Figure 3.7 Selected policy options for the design of market-based measures**

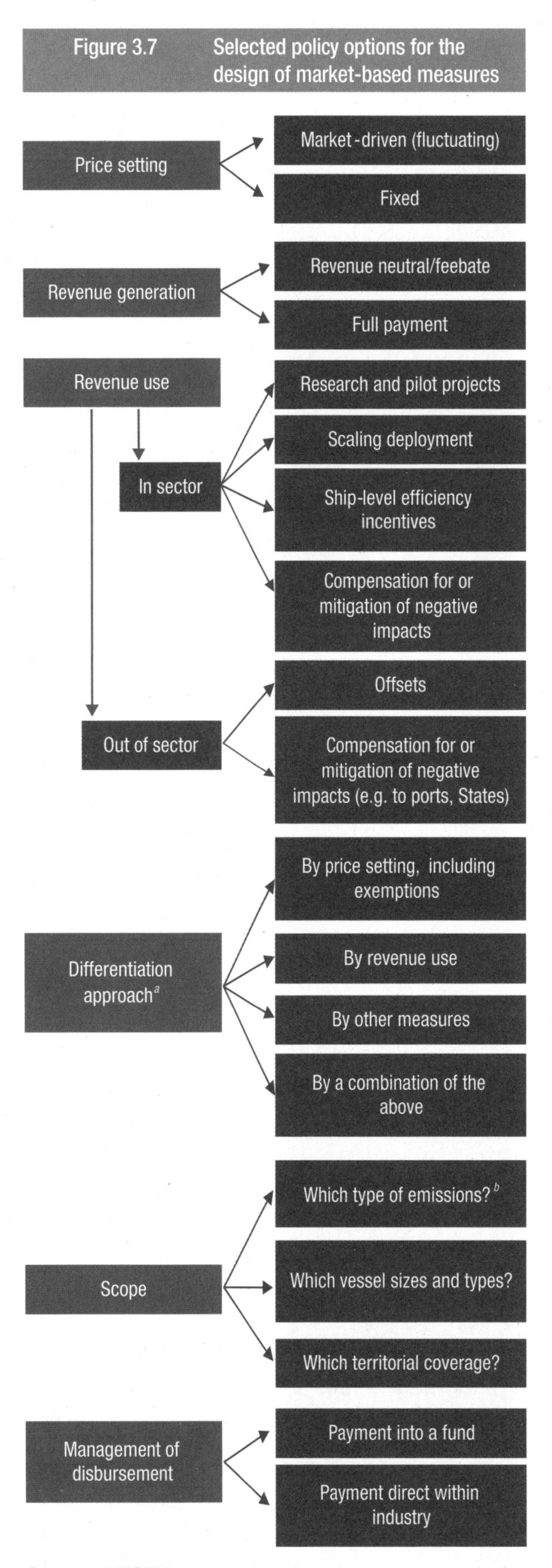


*Source:* UNCTAD secretariat, based on a categorization proposed by Tristan Smith, University College London.

[a] Common but differentiated responsibilities and respective capabilities.

[b] Only carbon dioxide or all greenhouse gas emissions.

could be charged, and the revenues distributed to the vessels emitting less than the benchmark (feebate). This would limit the amounts collected – thus alleviating the impact on transport costs and trade distortion and consequently the need for compensatory action, while continuing to provide a strong incentive to increase efficiency. Nevertheless, establishing a metric for the benchmark could prove to be complex.

Collecting revenue for all emissions instead of the balance to a benchmark could be less complex to implement at the policy level, and the challenge of establishing a metric for the benchmark may be avoided. Clearly, the revenue raised from all emissions would be higher, which in turn would provide more funds to support decarbonization in broader ways. A major disadvantage would be the stronger transport cost and trade distortion impact, given the higher amount of carbon allowances to be purchased.

#### *Revenue use and differentiation*

Revenues generated by the proposed market-based measures could be used by the maritime transport sector (in sector) to accelerate the development of clean and efficient technology. Revenues generated could be used to support research and pilot projects, scale up the deployment of relevant technologies and thus enable new technologies to reach economies of scale and become competitive. Funds could also be used to provide incentives for ships by distributing some revenues to vessels considered to be more efficient and to have a lighter carbon footprint. This can provide an incentive to shipowners and operators to further invest and implement relevant technologies and solutions. The funds could also be used outside the maritime transport sector (out of sector). Examples include using the funds as carbon offsets by financing greenhouse gas emissions reduction measures in other sectors that would compensate for shipping emissions. The funds could also be used to compensate or mitigate the negative impact of some greenhouse gas emissions reduction measures.

Any carbon-pricing instrument, however, should reflect the IMO principle of non-discrimination and no more favourable treatment between ships, as well as the principle of common but differentiated responsibilities and respective capabilities applied under the United Nations Framework Convention on Climate Change, including under the Paris Agreement. The guiding principles of the initial IMO strategy on the reduction of greenhouse gas emissions state that the strategy will be cognizant of both these approaches (IMO, 2018). The differentiation could be delivered by various means: The allowance price could be differentiated by ship type, ship size or route – with an exemption effectively representing a price of zero, and/or the revenue use be handled along the common but differentiated responsibilities and respective capabilities principle. In this variant, the revenue could be used to compensate for or mitigate negative impacts from the greenhouse gas emissions reduction scheme, such as an increase in transport costs. The revenue could be disbursed to States to absorb negative impacts on imports or exports, to shipowners or shipyards to build a clean fleet, to port and other transport infrastructure operators to improve efficiency and bring down transport costs at their respective level of the supply chain or to fuel suppliers to develop low-carbon fuels. All these options pose a risk of improper usage of funds and may create market distortion. On the other hand, funds could be directed to support investments in the transport systems of developing countries.

#### *Scope and enforcement*

In general, the scope of a greenhouse gas emissions-reduction scheme for shipping should cover various elements. For instance, should the scheme cover all greenhouse gas emissions or only carbon dioxide? Which vessel sizes and types should be considered? Should emissions from international sea transport be the only emissions included or should domestic shipping also be taken into account? Should the price be set per unit of fuel or per ton of carbon dioxide? In addition, a strong and reliable audit and enforcement system is required. Compliance could be checked by port State control by means of the bunker delivery note, the oil record book or the IMO data collection system.

## 2. The impact of carbon prices on freight rates

Assessing the effects of carbon-pricing schemes that may be adopted in maritime transport and understanding the potential implications for transport and trade requires further analytical work. Existing research should provide some relevant insights. In a survey conducted by Lloyd's Register and University Maritime Advisory Services (2018), some 75 per cent of shipowners agreed that a carbon price was needed, and that most would be willing to pay $50 per ton of carbon dioxide. The International Monetary Fund estimates that a carbon price higher than this, reaching $75 per ton by 2030, would reduce emissions in that year by about 15 per cent compared with a business as usual scenario and by about 11 per cent compared with 2008 levels (Parry et al., forthcoming). To reach the goal of 50 per cent or more by 2050, analysis carried out by University College London reveals that a carbon price of $100–$300 per ton of carbon dioxide would be necessary for the related technology to be competitive. This assumes no complementary policies other than those already in place and production of maritime fuels with electricity prices equivalent to some of the lowest prices today. The estimate is lower than previous analyses and takes into account the expected increase in fuel costs due to the global cap on sulphur content, which will take effect in 2020. The combustion of one ton of oil-based bunker fuel produces about three tons of carbon dioxide (IMO, 2008).

The impact of a universal carbon price on emissions from maritime transport on freight rates and transport costs would depend on several parameters, including market structure, trade routes and cargo type. According to Kosmas and Acciaro (2017), the carrier can pass on the additional cost to shippers in a demand-driven market, whereas this is less true in a supply-driven market. This is demonstrated by a comparison of market conditions in 2006–2007, characterized by high demand and elevated freight rates, and 2012–2013, when there was high overcapacity. If a hypothetical fuel levy had been introduced in 2006–2007, 48 per cent of the levy would have been borne by carriers, and 52 per cent by shippers. In the overcapacity situation of 2012–2013, it is estimated that 90.3 per cent would have been borne by carriers, and 9.7 per cent by shippers. However, the authors noted that operational fuel-efficiency practices such as slow steaming would also increase, lessening the amounts due for the levy.

Studies focusing on the impact of bunker fuel cost increases on freight rates provide some indication of the potential implications of a carbon price, including in the form of a fuel levy. UNCTAD estimated the correlation between fuel prices and maritime freight rates from 1993 to 2008 and concluded that freight rates were sensitive to changes in fuel price, with variations by market segment (UNCTAD, 2010). The analysis showed a price elasticity of 0.17 to 0.34 of container freight rates in response to Brent crude oil prices (a good proxy for bunker fuel prices) over the time period covered. Therefore, a 10 per cent increase in shipping fuel costs would lead to an increase of 1.7–3.4 per cent in container freight rates. In times of higher oil prices, such as between 2004 and 2008, the elasticity tended to be at the upper level of the range. Vivid Economics (2010) put forward an estimate for different types of cargo and found on average an elasticity of 0.37 for very large crude carriers, 0.25 for Panamax grain carriers, 0.96 for Capesize ore carriers and 0.11 for container ships.

Costs arising from carbon pricing are likely to be route specific, and their extent will be influenced by other factors that determine shipping rates and transport costs. These include distance, trade imbalances, features of the products shipped (low-value high-volume goods are particularly sensitive to fuel prices), availability of slow steaming as a shock absorber, efficiency of ships deployed (newer and larger vessels tend to be more efficient) and port characteristics (UNCTAD, 2015; Vivid Economics, 2010). In the future, the question of who has access to low-cost renewable energy sources for biomass- and electricity-based fuels will also play a role in terms of transport cost (Lloyd's Register and University Maritime Advisory Services, 2018).

International transport costs are a crucial determinant of a developing country's trade competitiveness and often represent a constraint to greater participation in international trade. For the least developed countries, transport costs represented 21 per cent of the value of imports in 2016, and 22 per cent, for small island developing States, as opposed to 11 per cent for developed economies (UNCTAD, 2017). While it is essential to meet greenhouse gas emissions reduction targets in maritime transport, it is also important to consider the special needs of the most vulnerable economies that face acute logistical challenges and high transport costs hindering their market access and driving up their transport costs and import expenditure. These economies include, in particular the least developed countries and small island developing States. Accounting for the varied conditions and the wide-ranging market structures will help ensure that any market-based measures introduced would not increase the import bill or undermine the potential of developing countries to participate in global value chains and trade. If, for example, small island developing States were to lose export competitiveness because of carbon costs, and could not substitute imports with local production, this would drive transport costs up even further due to empty returns (UNCTAD, 2010).

As ongoing research work and discussions on potential mitigation policies under IMO continue, the international community – carriers, shippers, policymakers and others – needs to further discuss and assess the various options available and promote the adoption of widely accepted solutions to ensure effective implementation. Delays in implementing a robust low-carbon trajectory will increase the time pressure and require a rapid reduction in emissions in future. This in turn may drive up costs, especially given the locked-in investments in the transport sector.

Besides a timely entry into force, another cornerstone of any future market-based measure adopted under the auspices of IMO relates to the design and structure of the measure. It should be flexible to allow adaptability to changing market trends and realities. Although projections are pointing to a positive outlook, how maritime transport demand will evolve over the next 30 years will be subject to a high degree of uncertainty, owing to the numerous downside risks and emerging trends that entail both challenges and opportunities for the maritime transport sector (see chapters 1, 2 and 5). Any forthcoming mitigation measures or underlying policy frameworks should therefore be flexible to adapt to a fast-changing operating and regulatory landscape, while ensuring a price signal that incentivizes investment and generates revenues. Such funds could be used as investments to reduce transport costs, especially in developing countries, where such costs can be prohibitive and often serve as a stronger barrier to trade than tariffs.

## E. OUTLOOK AND POLICY CONSIDERATIONS

In 2017, freight rate levels improved significantly and, with the exception of the tanker market, reached levels above the performances recorded in 2016. The recovery in rates reflected a strengthening of global demand, combined with a deceleration in fleet capacity growth. Together, these factors resulted in overall healthier market conditions. Despite the marked improvement, the sustainability of the recovery remains at risk. This is due to the high volatility and relatively low levels of freight rates, as well as the potentially dampening effect of downside risks weighing on the demand side and the risk of inadequate supply capacity management.

UNCTAD projects global containerized trade to expand at a compound annual growth rate of 6.4 per cent in 2018 and 6.0 per cent between 2018 and 2023 (see chapter 1). Growth of global ship supply capacity is expected to remain fairly moderate over the next few years. World fleet capacity is projected to rise by 3 per cent in 2018; a growing share of additional capacity will be attributed to larger-size vessels (see chapter 2). Based on these projections, market balance should continue improving in the short term. Freight rates may benefit accordingly, although supply-side capacity management and deployment remain crucial, given the ongoing delivery of and new orders for mega vessels.

However, it is unlikely that in 2018 the industry will report the healthy profit estimated in 2017: despite the improvements observed in freight rates, the latest increase in fuel prices might affect the profitability of shipping lines.

The trend toward liner consolidation with mergers and acquisitions and realignment of the alliances among carriers continues in line with market conditions in 2018. Companies are likely to continue to seek opportunities to increase their market shares, improve efficiency and deal with intensifying competition and persistent oversupply. Consolidation through alliances would allow shipping companies to pool their resources and increase efficiencies. Larger shipping lines would aim to rationalize their resources in an alliance, whereas smaller lines would be able to enjoy the extended service coverage without having to invest in a larger fleet (Freight Hub, 2017). However, those that are not part of an alliance may be at a competitive disadvantage, as they may not be able achieve the cost efficiencies required to compete with members of an alliance. On the other hand, niche carriers that have a specific focus on a market or region and do not compete with larger firms on the main trade lanes may not feel the threat (World Maritime News, 2017).

The impact of consolidation has yet to be fully understood. While outright negative impacts on trade and costs have not been reported, there are remaining concerns about the impact of growing market concentration on competition and the level playing field. However, it may be argued that larger lines can offer more services and make relevant investments including in technology, which in turn could drive down costs through greater economies of scale and higher levels of efficiency. Some experts say that the larger the line, the easier it is to change the network offering which translates into more flexibility and adaptability to changing market conditions (The Maritime Post, 2018).

Competition authorities and regulators as well as transport analysts and international entities such as UNCTAD should remain vigilant by continuing to monitor consolidation activity and assess the market concentration level and the potential for market power abuse by large shipping lines and the related impact on smaller players and potential implications in terms of freight rates and other costs to shippers and trade. An analysis of mergers and alliances should consider not only the effects of price competition, but also the variety and quality of services provided to shippers. Competition authorities should take into account the effects on the range and quality of services, frequency of ships, range of ports serviced, reliability of schedules and efficiency. In this respect, the seventeenth session of the Intergovernmental Group of Experts on Competition Law and Policy included a round-table discussion on challenges in competition and regulation faced by developing countries in the maritime transport sector. This provided a timely opportunity to bring together representatives of competition authorities and other stakeholders from the sector to reflect upon some of these concerns and assess the extent and potential implications for competition, shipping and seaborne trade, as well as the role of competition law and policy in addressing these concerns (UNCTAD, 2018).

With regard to the prospects of the various market segments, the dry bulk market is set to further improve in 2018, supported by projected growth (5.2 per cent compound annual growth rate in 2018 and 4.9 per cent between 2018 and 2023) and the more subdued projected growth (3 per cent) in the bulk carrier fleet. Together, these improvements to the fundamental balance will sustain positive dry bulk shipping rates in 2018. That said, downside risks remain, such as the trade policy risks identified in chapter 1, in particular the impact of United States tariffs on steel and aluminium from Canada, Mexico and the European Union. Tanker trade volumes are also projected to increase, although at a slightly slower pace than other market segments. However, overcapacity may continue to depress the conditions in the tanker shipping freight market.

Of particular relevance for transport costs and shippers' expenditure on sea carriage are the ongoing developments in IMO that might result in market-based measures aimed at reducing carbon emissions from shipping as part of a comprehensive package of mitigation actions. As research work and discussions on potential mitigation policies to be adopted under the auspices of IMO continue, the international community – industry, shippers, trade, policymakers and others – needs to further discuss and assess the various options available and promote the adoption of widely accepted solutions to ensure effective implementation. Delays in implementing a robust low-carbon trajectory will increase the time pressure and require a rapid reduction in emissions. This in turn, may drive up costs, especially given locked-in investments. Besides a timely entry into force, another cornerstone of any future market-based measures adopted under the auspices of IMO relates to design. The latter should be flexible to allow adaptability to market developments. Although projections tend to be positive, the issue of how global and local maritime transport demand will evolve over the next 30 years is subject to a high degree of uncertainty, driven by a wide range of prevalent downside risks and emerging trends that will bring challenges and opportunities for the maritime transport sector (see chapters 1, 2 and 5). Any mitigation policy should therefore be flexible to adapt to fast-changing operating and regulatory landscapes, while ensuring a price signal that incentivizes investment and generates revenues. The latter could be used as investments to reduce transport costs, especially in developing countries, where transport costs are generally more prohibitive than the world average. In this respect, a focus on the special needs of the least developed countries and small island developing States is warranted.

# REFERENCES

A. P. Moller–Maersk (2018). *2017 Annual Report*. Copenhagen. Available at http://investor.maersk.com/static-files/250c3398-7850-4c00-8afe-4dbd874e2a85.

Barry Rogliano Salles (2018). Annual review 2018: Shipping and shipbuilding markets. Available at: https://it4v7.interactiv-doc.fr/html/brsgroup2018annualreview_pdf_668.

Carbon Pricing Leadership Coalition (2018). What is carbon pricing? Available at www.carbonpricingleadership.org/what/.

Clarksons Research (2018a). *Container Intelligence Quarterly*. First quarter 2018.

Clarksons Research (2018b). *Dry Bulk Trade Outlook*. Volume 23. No. 1. January.

Clarksons Research (2018c). *Shipping Review and Outlook*. Spring.

Clarksons Research (2018d). Shipping Intelligence Network – Timeseries.

CMA CGM (2018a). 2017 annual financial results. Available at www.cma-cgm.com/news/1973/2017-annual-financial-results-cma-cgm-pursues-its-development-strategy-and-once-again-delivers-strong-operating-results-outperforming-its-industry.

CMA CGM (2018b). Consolidated financial statements: Year ended 31 December 2017. Available at www.cma-cgm.com/static/Finance/PDFFinancialRelease/2017%20-%20Annual%20Consolidated%20Accounts.pdf.

Drewry (2018). *Container Forecaster*. First quarter. March.

Freight Hub (2017). Shipping alliances: What do they do and what does it mean? Available at https://freighthub.com/en/blog/shipping-alliances-mean/.

Hapag-Lloyd (2017). Hapag-Lloyd successfully completes integration with UASC [United Arab Shipping Company]. 30 November. Available at www.hapag-lloyd.com/en/press/releases/2017/11/hapag-lloyd-successfully-completes-integration-with-uasc.html.

Hapag-Lloyd (2018). *Annual Report 2017*. Hapag-Lloyd Corporate Communications, Hamburg.

Hellenic Shipping News (2017). Demolition trends: Global fleet ups its game. 29 July. Available at: www.hellenicshippingnews.com/demolition-trends-global-fleet-ups-its-game/.

Hellenic Shipping News (2018). Tanker freight rates at below operating expenses despite seasonality factor. 5 February. Available at www.hellenicshippingnews.com/tanker-freight-rates-at-below-operating-expenses-despite-seasonality-factor/.

IMO (2008). Marine Environment Protection Committee. Report of the Drafting Group on Amendments to MARPOL [International Convention for the Prevention of Pollution from Ships] annex VI and the NOx [Nitrogen Oxides] Technical Code. MEPC 58/WP.9. London. 8 October.

IMO (2018). Adoption of the initial IMO strategy on reduction of greenhouse gas emissions from ships and existing IMO activity related to reducing greenhouse gas emissions in the shipping sector.

JOC.com (2017). Ship charter rates surge on demand, alliance capacity. 11 April.

Kosmas V and Acciaro M (2017). Bunker levy schemes for greenhouse gas emission reduction in international shipping. *Transportation Research Part D: Transport and Environment.* 57:195–206.

Lloyd's Loading List (2017). Container lines make losses of $3.5bn in 2016. 3 April. Available at www.lloydsloadinglist.com/freight-directory/news/Container-lines-make-losses-of-3.5bn-in-2016/68969.htm#.WwqCGCC-mMo.

Lloyd's Register and University Maritime Advisory Services (2018). Zero-emission vessels 2030. How do we get there? Low Carbon Pathways 2050 Series. Available at www.lr.org/en/insights/articles/zev-report-article/.

MDS Transmodal (2018). Container ship databank. June.

Organization for Economic Cooperation and Development (2018). Emission trading systems. Available at www.oecd.org/env/tools-evaluation/emissiontradingsystems.htm.

Parry I, Heine D, Kizzier K and Smith T (forthcoming). Carbon taxation for international maritime fuels: Assessing the options. Working paper. International Monetary Fund. Washington, D.C.

Psaraftis HN (2012). Market-based measures for greenhouse gas emissions from ships: A review. *World Maritime University Journal of Maritime Affairs*. 11(2):211–232.

The Loadstar (2018). Healthier new year for container charter market, but owners still have concerns. 8 January. Available at https://theloadstar.co.uk/healthier-new-year-container-charter-market-owners-still-concerns/#.

The Maritime Post (2018). Top 10 shipping lines control almost 90% of the deep sea market. 26 February. Available at www.themaritimepost.com/top-10-shipping-lines-control-almost-90-deep-sea-market/.

UNCTAD (2010). *Oil Prices and Maritime Freight Rates: An Empirical Investigation*. UNCTAD/DTL/ TLB/2009/2.

UNCTAD (2015). *Review of Maritime Transport 2015* (United Nations publication, Sales No. E.15.II.D.6, New York and Geneva).

UNCTAD (2017). *Review of Maritime Transport 2017* (United Nations publication, Sales No. E.17.II.D.10, New York and Geneva).

UNCTAD (2018). Challenges faced by developing countries in competition and regulation in the maritime transport sector. TD/B/C.I/CLP/49. Geneva. 2 May.

Vivid Economics (2010). Assessment of the economic impact of market-based measures. Final report. Prepared for the IMO Expert Group on Market-based Measures.

World Maritime News (2017). Moody's: Carriers' consolidation will continue into 2018. Available at: https://worldmaritimenews.com/archives/237994/moodys-carriers-consolidation-will-continue-into-2018/.

# ENDNOTES

1. Three shipping alliances were formed in 2018: 2M, the Ocean Alliance and "The" Alliance. The first, 2M, is composed of the Mediterranean Shipping Company and Maersk, which acquired Hamburg Süd. (Hyundai Merchant Marine signed a strategic cooperation agreement with the 2M partners.) The second, the Ocean Alliance, brought together three shipping lines, CMA CGM, which acquired American President Lines and Mercosul Line; China Cosco Shipping, which acquired Orient Overseas Container Line; and Evergreen. The third, "The" Alliance, was born of a merger between Hapag-Lloyd, Yang Ming and Ocean Network Express (the latter is also known as "ONE", a joint venture established between Nippon Yusen Kabushiki Kaisha, Mitsui Osaka Shosen Kaisha Lines and Kawasaki Kisen Kaisha in April 2018).
2. This section benefits from comments provided during an informal workshop on market-based measures in maritime transport organized by the Carbon Pricing Leadership Coalition in Cologne, Germany, on 8 and 9 May 2018.
3. A summary of earlier discussions and/or proposals on market-based measures at IMO can be found in previous editions of the *Review on Maritime Transport*: 2010 (pp. 119–123), 2011 (pp. 118 and 119), 2012 (pp. 99–101) and 2013 (p. 108).
4. The emissions figures for 2008 and for the 2030 projection are based on different sources, which might slightly influence the relative reduction figure.

In 2017, global port activity and cargo handling of containerized and bulk cargo expanded rapidly, following two years of weak performance. This expansion was in line with positive trends in the world economy and seaborne trade. Global container terminals boasted an increase in volume of about 6 per cent during the year, up from 2.1 per cent in 2016. World container port throughput stood at 752 million TEUs, reflecting an additional 42.3 million TEUs in 2017, an amount comparable to the port throughput of Shanghai, the world's busiest port.

While overall prospects for global port activity remain bright, preliminary figures point to decelerated growth in port volumes for 2018, as the growth impetus of 2017, marked by cyclical recovery and supply chain restocking factors, peters out. In addition, downside risks weighing on global shipping, such as trade policy risks, geopolitical factors and structural shifts in economies such as China, also portend a decline in port activity.

Today's port-operating landscape is characterized by heightened port competition, especially in the container market segment, where decisions by shipping alliances regarding capacity deployed, ports of call and network structure can determine the fate of a container port terminal. The framework is also being influenced by wide-ranging economic, policy and technological drivers of which digitalization is key. More than ever, ports and terminals around the world need to re-evaluate their role in global maritime logistics and prepare to embrace digitalization-driven innovations and technologies, which hold significant transformational potential.

Strategic liner shipping alliances and vessel upsizing have made the relationship between container lines and ports more complex and triggered new dynamics, whereby shipping lines have stronger bargaining power and influence. The impact of liner market concentration and alliance deployment on the port–carrier relationship will need to be monitored and assessed. Areas of focus include the impact on the selection of ports of call, the configuration of liner shipping networks, the distribution of costs and benefits between container shipping and ports, and approaches to container terminal concessions, as shipping lines often have stakes in terminal operations.

Enhancing port and terminal performance in all market segments is increasingly recognized as critical for port planning, investment and strategic positioning, as well as for meeting globally established sustainability benchmarks and objectives such as the Sustainable Development Goals. Ports and their stakeholders, including operators, users and Governments, should collaborate to identify and enable key levers for improving port productivity, profitability and operational efficiencies.

# PORTS IN 2017

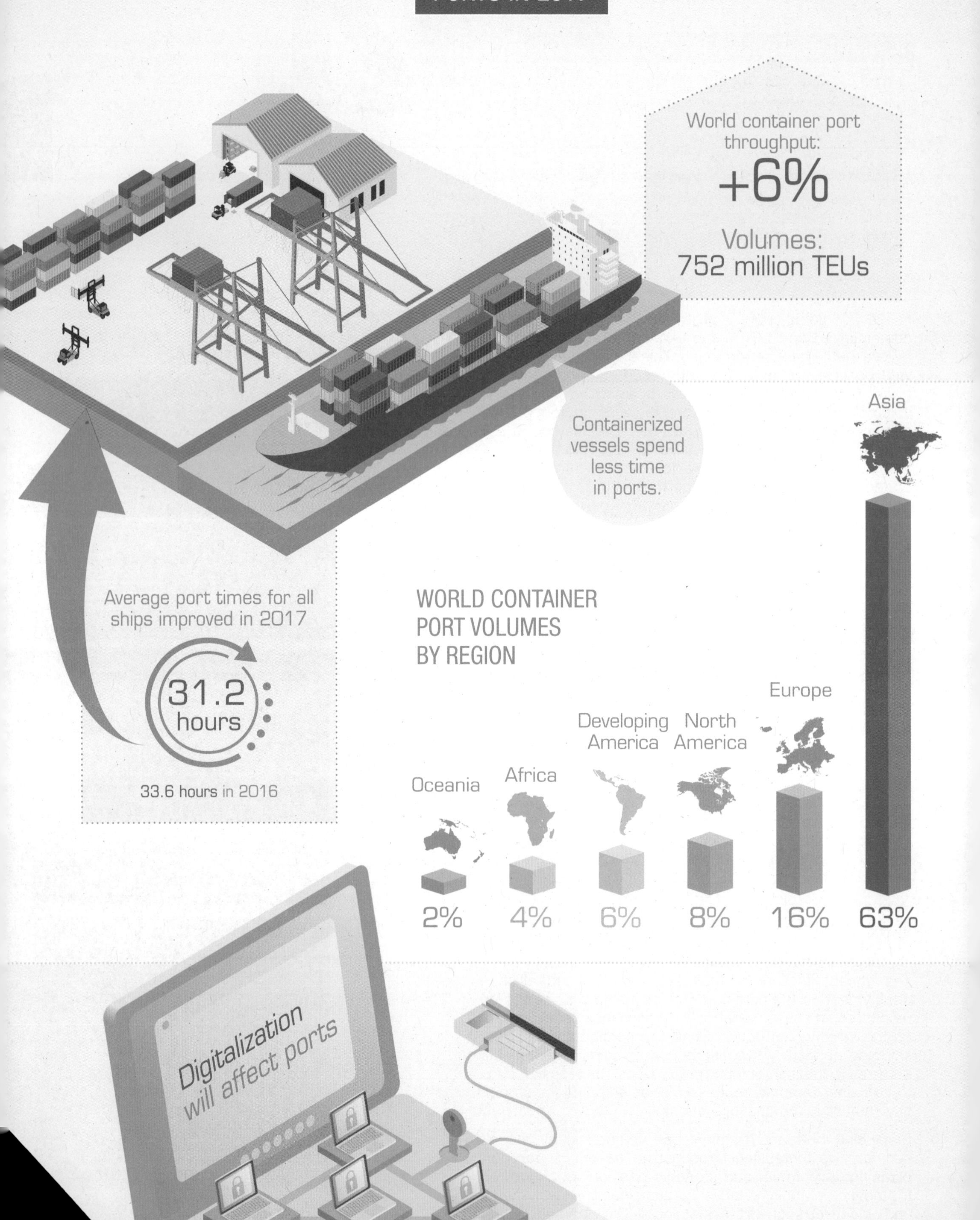

## A. OVERALL TRENDS IN GLOBAL PORTS

As key players in international trade and logistics and critical nodes in global supply chains, seaports continue to underpin globalized production processes, market access and effective integration in the global economy. World seaports are principal infrastructural assets that service shipping and trade, and their performance is largely determined by developments in the world economy and trade. Cargo-handling activity and throughput in global ports, which reflected a recovery in the global economy and a rebound in trade volumes that boosted shipping demand and seaborne trade in 2017, showed overall improvement and promising trends.

Since over 80 per cent of world merchandise trade in volume terms is handled by ports worldwide and nearly two thirds of this trade is loaded and unloaded in the ports of developing countries, the strategic importance of well-functioning and efficient ports for growth and development cannot be overemphasized. Global ports cater to ships and cargo across various stages of port-handling operations, starting with the shoreside, to the berth, the yard and the landside. Therefore, enhancing port efficiency throughout the various cargo- and vessel-handling phases is crucial for overall efficiency and to ensure that gains achieved by one segment of the maritime logistics chain are not cancelled out by inefficiencies arising elsewhere in the process.

Ports are at the intersection of many developments. They benefited from a global recovery in 2017 that remains nevertheless fragile, owing to ongoing downside risks. They also face challenges arising from the changing dynamics in the liner shipping market, the need to embrace technological advances brought about by digitalization, the requirement to comply with a heightened global sustainability agenda and the imperative of remaining competitive and responding to the demands of the world economy and trade.

### 1. Improvements in global port cargo throughput

A widely used indicator providing insights into the functioning of ports and their ability to attract business is volumes handled by ports. As cargo flows are largely determined by changes in demand, port volumes help take the pulse of the world economy and inform about potential transport infrastructure needs and investment requirements. As such, port cargo throughput, including all cargo types, can serve as a leading economic indicator. While data for global port throughput in 2017 was not available at the time of writing, a look at data for 2016 indicates the scale of overall port-handling activity. Cargo throughput (all cargo types, including containerized and bulk commodities) at world major ports was estimated at over 15 billion tons in 2016, following an increase of 2.1 per cent over 2015 (Shanghai International Shipping Institute, 2016).

A study describing the performance of leading global ports between 2011 and 2016 found that bulk-handling terminals captured most of the expansion gains of all ports, including container- and bulk-handling ports (Fairplay, 2017a). Almost all leading ports recorded a volume increase, except Shanghai, where the amount of cargo handled declined over the review period. With 485 million tons handled in 2016, Port Hedland, Australia saw rapid growth during the same period, followed by the Chinese ports of Ningbo-Zhoushan, Caofeidian, Tangshan and Suzhou. The top 20 global ports included only three ports outside Asia: the ports of Hedland, Rotterdam and South Louisiana. Compared with other ports on the list, cargo handled at the port of Rotterdam expanded at a slower rate between 2011 and 2016, owing to a relative decline in bulk commodity volumes handled. Overall, and despite their predominance, port volumes in China are said to be increasingly affected by the country's gradual transition towards a more service- and consumption-oriented economy. In Singapore, port volumes between 2011 and 2016 increased, and the first liquefied natural gas bunkering terminal was opened in 2017.

Preliminary analysis suggests that port volumes increased in 2017 reflecting, to a large extent, global economic recovery and growth in seaborne trade (see chapter 1). Estimates indicate that volumes handled in the top 20 global ports increased by 5 per cent to 9.4 billion tons in 2017, compared with 8.9 billion tons in 2016 (Shanghai International Shipping Institute, 2017).

Table 4.1. provides a list of leading global ports, measured by total tons of all cargo handled. Among the top 10 ports, 8 were in Asia, mainly from China. Ningbo-Zhoushan ranked first, with total volumes handled surpassing the 1 billion ton mark for the first time. Aside from Tianjin, which saw an 8.4 per cent drop in volumes, all ports on the list recorded volume increases in 2017. Reduced volumes in Tianjin may reflect the delayed effect of the industrial accident that occurred in 2015 and involved two explosions in the port's storage and handling of hazardous materials facilities. It may also reflect government restrictions on the use of tracks for the carriage of coal. With regard to Shanghai, the continued rebalancing of the Chinese economy towards domestic consumption and services was a major factor in the port's ranking.

Global port activity, which mirrored global economic recovery in 2017, improved across all regions, albeit with some variations. Existing data highlight the positive performance of ports in Europe and the United States, with volumes handled increasing at an annual rate of 4.9 per cent and 7 per cent,

**Table 4.1 Global top 20 ports by cargo throughput, 2016–2017**
(Million tons and annual percentage change)

| Rank | Port | Cargo throughput | | Percentage change |
|---|---|---|---|---|
| 2017 | | 2016 | 2017 | 2017–2016 |
| 1 | Ningbo-Zhoushan | 918 | 1 007 | 9,7 |
| 2 | Shanghai | 700 | 706 | 0,8 |
| 3 | Singapore | 593 | 626 | 5,5 |
| 4 | Suzhou | 574 | 608 | 5,9 |
| 5 | Guangzhou | 522 | 566 | 8,5 |
| 6 | Tangshan | 516 | 565 | 9,6 |
| 7 | Qingdao | 501 | 508 | 1,4 |
| 8 | Port Hedland | 485 | 505 | 4,3 |
| 9 | Tianjin | 549 | 503 | -8,4 |
| 10 | Rotterdam | 461 | 467 | 1,3 |
| 11 | Dalian | 429 | 451 | 5,2 |
| 12 | Busan | 362 | 401 | 10,5 |
| 13 | Yingkou | 347 | 363 | 4,4 |
| 14 | Rizhao | 351 | 360 | 2,7 |
| 15 | South Louisiana | 295 | 308 | 4,4 |
| 16 | Gwangyang | 283 | 292 | 3,1 |
| 17 | Yantai | 265 | 286 | 7,6 |
| 18 | Hong Kong SAR | 257 | 282 | 9,7 |
| 19 | Zhanjiang | 255 | 282 | 10,3 |
| 20 | Huanghua | 245 | 270 | 10,0 |
| | **Total** | **8 907** | **9 354** | **5,0** |

*Source:* Shanghai International Shipping Institute, 2017.
*Note:* Figures cover all cargo types.
*Abbreviation:* SAR, Special Administrative Region.

respectively. Reflecting Asia's position as the main source of world shipping demand and the influence of China, port volumes handled at Asian ports increased by 7.2 per cent in 2017. Main ports in China handled 12.6 billion tons, an increase of 6.9 per cent over 2016. Ports in the Republic of Korea handled 1.57 billion tons, a 4.1 per cent improvement over 2016. Port volumes in Africa rose by 3.5 per cent, compared with 2016, reflecting overall improved economic conditions, a recovery in commodity export earnings and higher import demand in the region. Volumes handled at major ports in Australia expanded at the slow pace of 2.3 per cent in 2017, as port activity was affected by *Hurricane Debbie*. In particular, the hurricane undermined the performance of the port of Hay Point, the largest coal port in Australia.

## 2. Tracking and measuring port performance

Global trade, supply chains, production processes and countries' economic integration are heavily dependent on efficient port systems and supporting logistics. It is therefore becoming increasingly important to monitor and measure the operational, financial, economic, social and environmental performance of ports.

In 2013, the Port Management Programme of the UNCTAD Train for Trade Programme developed a port performance measurement component (see box 4.1). This work culminated in the adoption of 26 indicators across six areas: finance, human resources, gender, vessel operations, cargo operations and environment (UNCTAD, 2016). The main objective was to provide members of the Programme's port network with a useful instrument that would benchmark performance and carry out port and regional comparisons. Ports in the network involved in port performance measurement were landlord ports, full service ports, tool ports and mixed ports (figure 4.1). The port performance measurement system adopted under the Programme draws largely on the balance scorecard concept (table 4.2).

Results achieved between 2010 and 2017 are summarized in figures 4.2 to 4.6. When comparing port performance, the standard caveat is that ports are difficult to compare, with many context variables to consider. The scorecard describes the data profile for the 48 reporting ports since 2010 in terms of data set metrics, port size, modal mix, governance, market and regulatory structures. The indicators are sourced from wide-ranging ports, 66 per cent of which have annual volumes below 10 million tons.

Results presented in figures 4.2 to 4.6 reflect data provided by the reporting countries and port entities that are members of the network only. They should not

**Figure 4.1** **Port models of the Port Management Programme port network, 2016** (Share in percentage)

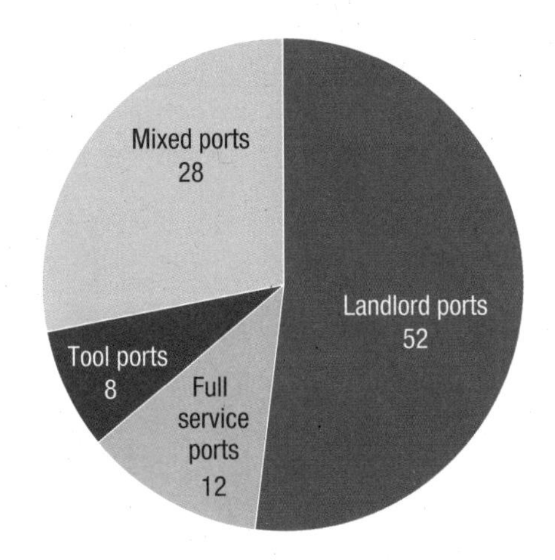


*Source:* UNCTAD, 2016.

be generalized or interpreted as reflecting all ports in the four regions defined under this scheme. Benchmarking has been developed for Asia, Africa, Europe and developing America. The global average is provided for all port networks of the Programme – French-, English-, Spanish- and Portuguese-speaking – reporting over a period of eight years and representing a total of 48 port entities from 24 countries.

Profit levels can vary considerably between ports, depending on the accounting treatment, capital reward structure and profit definition used in the indicator. Operating profit margins are considered the best level to make cross-country and time comparisons, given their composition. Therefore, the indicator is focused on the trading and management performance of the port entity. There are some outliers in the data, including a loss-making entity for one period. However, over time, the mean value has remained robust, ranging between 35 per cent and 45 per cent.

**Table 4.2** **Port performance scorecard indicators**

| Categories | | Port entity indicators | Number values | Mean in percentage (2010–2017) |
|---|---|---|---|---|
| Finance | 1 | EBITDA/revenue (operating margin) | 126 | 39,30 |
| | 2 | Vessel dues/revenue | 135 | 15,90 |
| | 3 | Cargo dues/revenue | 120 | 34,20 |
| | 4 | Rents/revenue | 117 | 10,10 |
| | 5 | Labour/revenue | 106 | 24,80 |
| | 6 | Fees and the like/revenue | 114 | 18,10 |
| Human resources | 7 | Tons per employee | 134 | 54 854 |
| | 8 | Revenue per employee | 128 | $235 471 |
| | 9 | EBITDA per employee | 107 | $119 711 |
| | 10 | Labour costs per employee | 89 | $42 515 |
| | 11 | Training costs/wages | 101 | 1,30 |
| Gender | 12 | Female participation rate, global | 54 | 15,70 |
| | 12,1 | Female participation rate, management | 53 | 30,90 |
| | 12,2 | Female participation rate, operations | 39 | 12,30 |
| | 12,3 | Female participation rate, cargo handling | 29 | 5,30 |
| | 12,4 | Female participation rate, other employees | 8 | 32,00 |
| | 12,5 | Female participation rate, management plus operations | 119 | 19,60 |
| Vessel operations | 13 | Average waiting time | 129 | 15 hours |
| | 14 | Average gross tonnage per vessel | 165 | 17 114 |
| | 15,1 | Oil tanker arrivals, average | 28 | 10,80 |
| | 15,2 | Bulk carrier arrivals, average | 28 | 11,20 |
| | 15,3 | Container ship arrivals, average | 28 | 40,30 |
| | 15,4 | Cruise ship arrivals, average | 29 | 1,80 |
| | 15,5 | General cargo ship arrivals, average | 28 | 16,50 |
| | 15,6 | Other ship arrivals, average | 27 | 19,10 |
| Cargo operations | 16 | Average tonnage per arrival (all) | 156 | 6 993 |
| | 17 | Tons per working hour, dry or solid bulk | 91 | 402 |
| | 18 | Boxes per hour, containers | 120 | 29 |
| | 19 | TEU dwell time, in days | 73 | 6 |
| | 20 | Tons per hour, liquid bulk | 46 | 299 |
| | 21 | Tons per hectare (all) | 130 | 131 553 |
| | 22 | Tons per berth metre (all) | 143 | 4 257 |
| | 23 | Total passengers on ferries | 18 | 811 744 |
| | 24 | Total passengers on cruise ships | 20 | 89 929 |
| Environment | 25 | Investment in environmental projects/total CAPEX | 10 | 0,90 |
| | 26 | Environmental expenditures/revenue | 17 | 0,30 |

*Source:* UNCTAD, 2016.
*Note:* Number of values is a product of ports providing data for the variable by the number of years reporting.
*Abbreviations:* CAPEX, capital expenditure; EBITDA, earnings before interest, taxes, depreciation and amortization.

Figure 4.2 Financial indicators, 2010–2017
(Share in percentage)

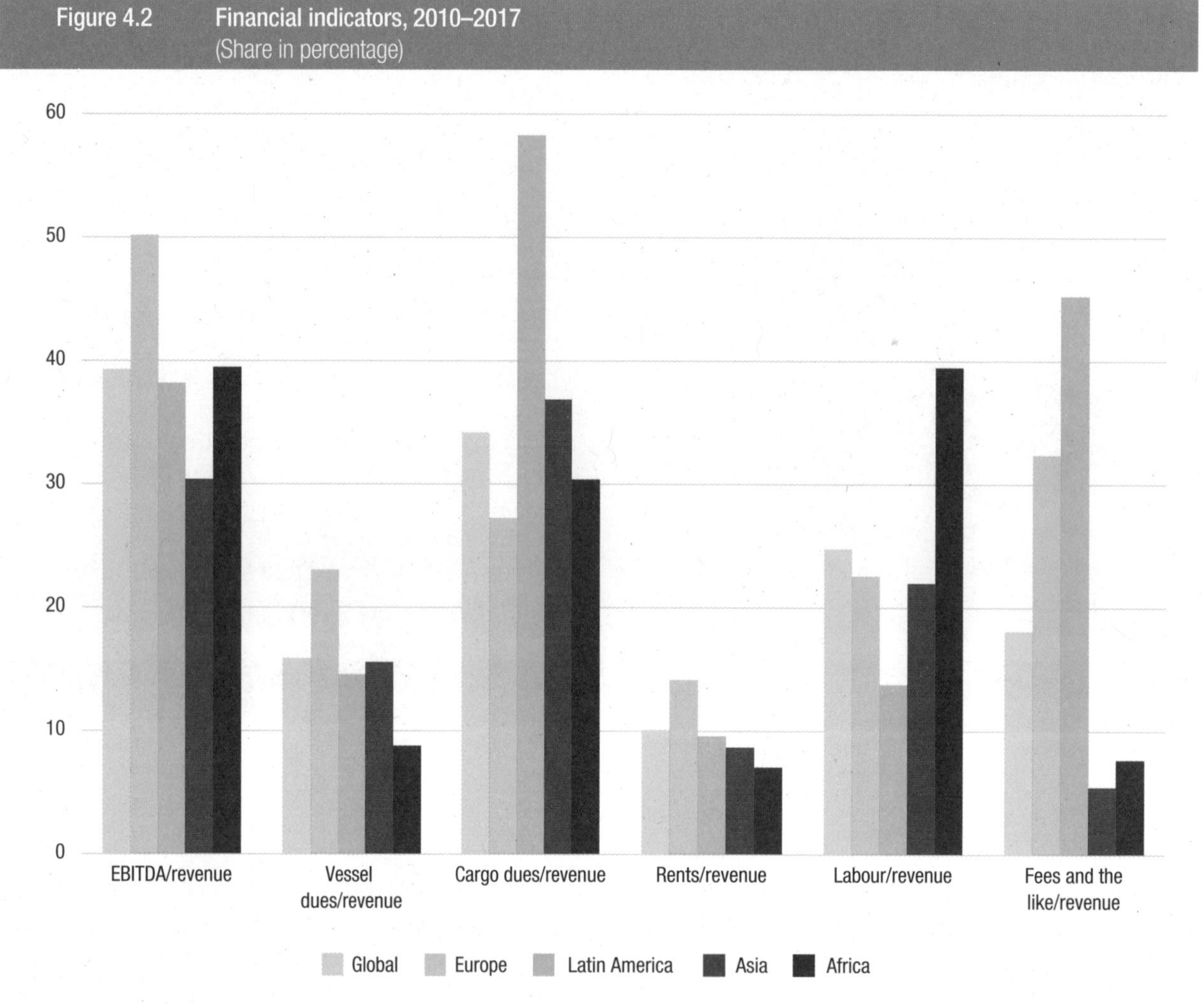


*Source:* UNCTAD, 2016.
*Abbreviation:* EBITDA, earnings before interest, taxes, depreciation and amortization.

It is useful to consider port dues for cargo and vessels together. The regional differences are less for the gross port dues (cargo plus vessels) proportion of revenue. Total revenue when averaged across volumes suggests that just over $4 is earned by a port entity on each ton of cargo.

Rent is a traditional source of independent income for ports. The clustering of the data in figure 4.2 is consistent with previous reporting. When contrasted with a concession or fee variable, it varies significantly across the network. There is a shift towards concessions to the private sector but thus far it has not necessarily implied a move away from leasing. It remains unclear whether this is due to concessions being added to a lease rather than replacing a lease.

Data in figure 4.3 are a significant addition to the scorecard and chart the changing gender balance across port authorities in the data set. There is a clear distinction between categories of employees across traditional lines that has yet to reflect the technological shift in working methods and skill sets on the quays. The data suggest that Africa is an outlier characterized by a high average payroll cost as a proportion of revenue. It remains unclear whether this could be attributed to lower revenue levels or higher staffing levels. The average wage is estimated at $47,000, with a large range of values. It is a number that requires considerable nuance and comparison with local economic indicators that will be examined in future port performance conferences.

Reflecting the growing importance of containerized trade and the role of containers in multimodal transport, container ship arrivals represented 36 per cent of all arrivals during the review period. Given that 48 port entities located in 24 countries provided data entries in the system for almost all 26 indicators, data points are above 100. This enhances the robustness of the statistical results, which can, nevertheless, be further improved through additional port reporting. Work aimed at interpreting the results has been initiated, including the use of a five-year moving average for analysis. There remains the question, however, of how insights generated from this work can be further leveraged to support informed strategic planning and decisions relating to ports.

**Figure 4.3** **Female participation rate, by area of activity, 2010–2017**
(Percentage)

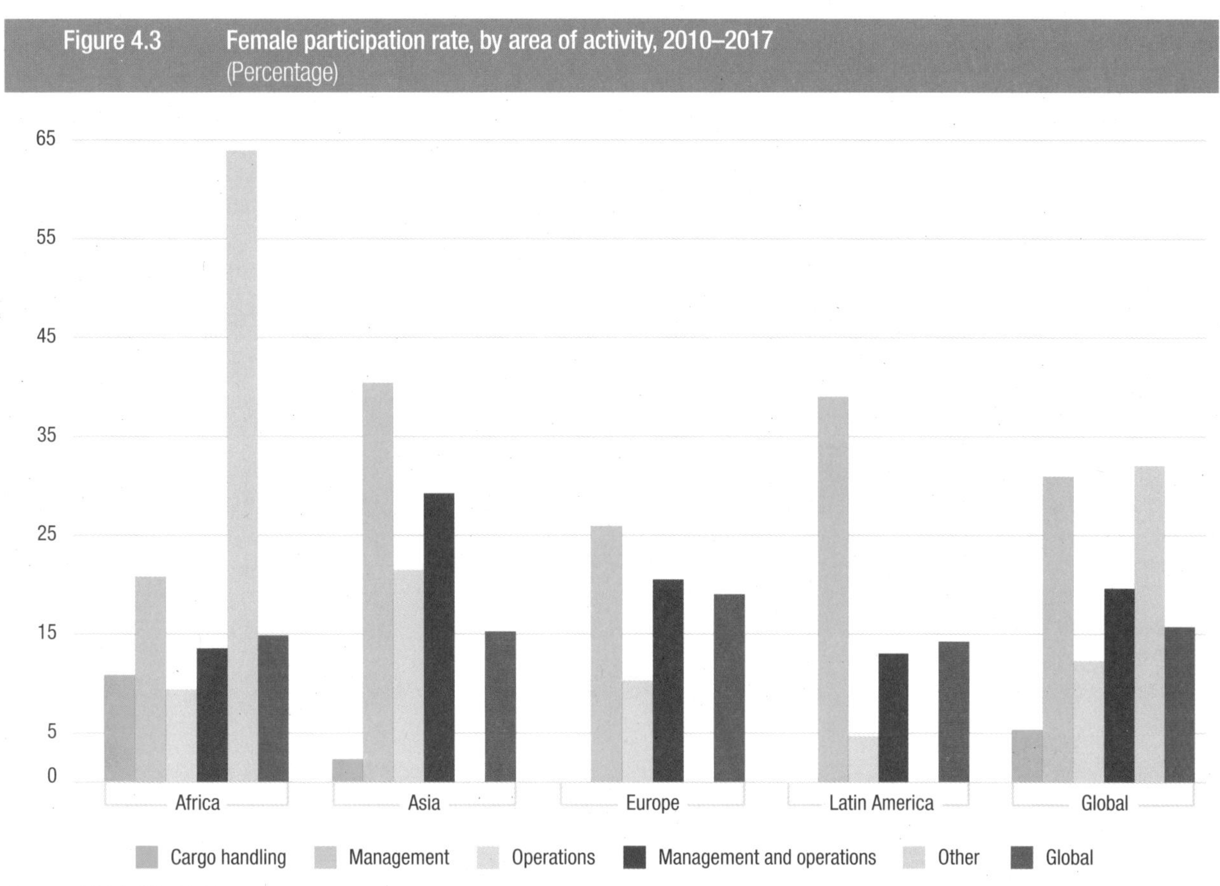


*Source:* UNCTAD, 2016.
*Note:* The female participation rate includes a five-year moving average.

**Figure 4.4** **Average arrivals by type of vessel, 2010–2017**
(Share in percentage)

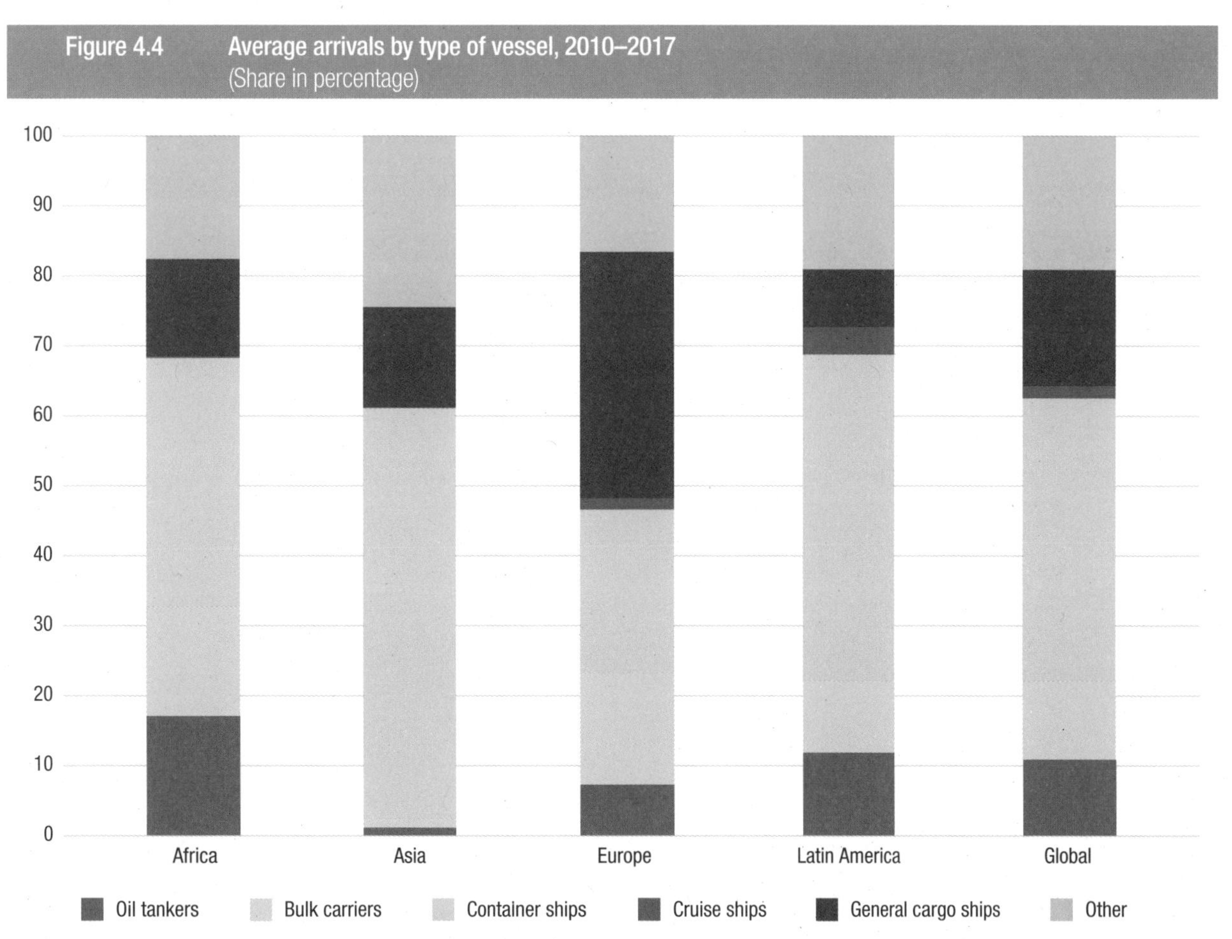


*Source:* UNCTAD, 2016.

Figure 4.5 Dry and liquid bulk cargo operations, 2010–2017
(Tons per working hour)

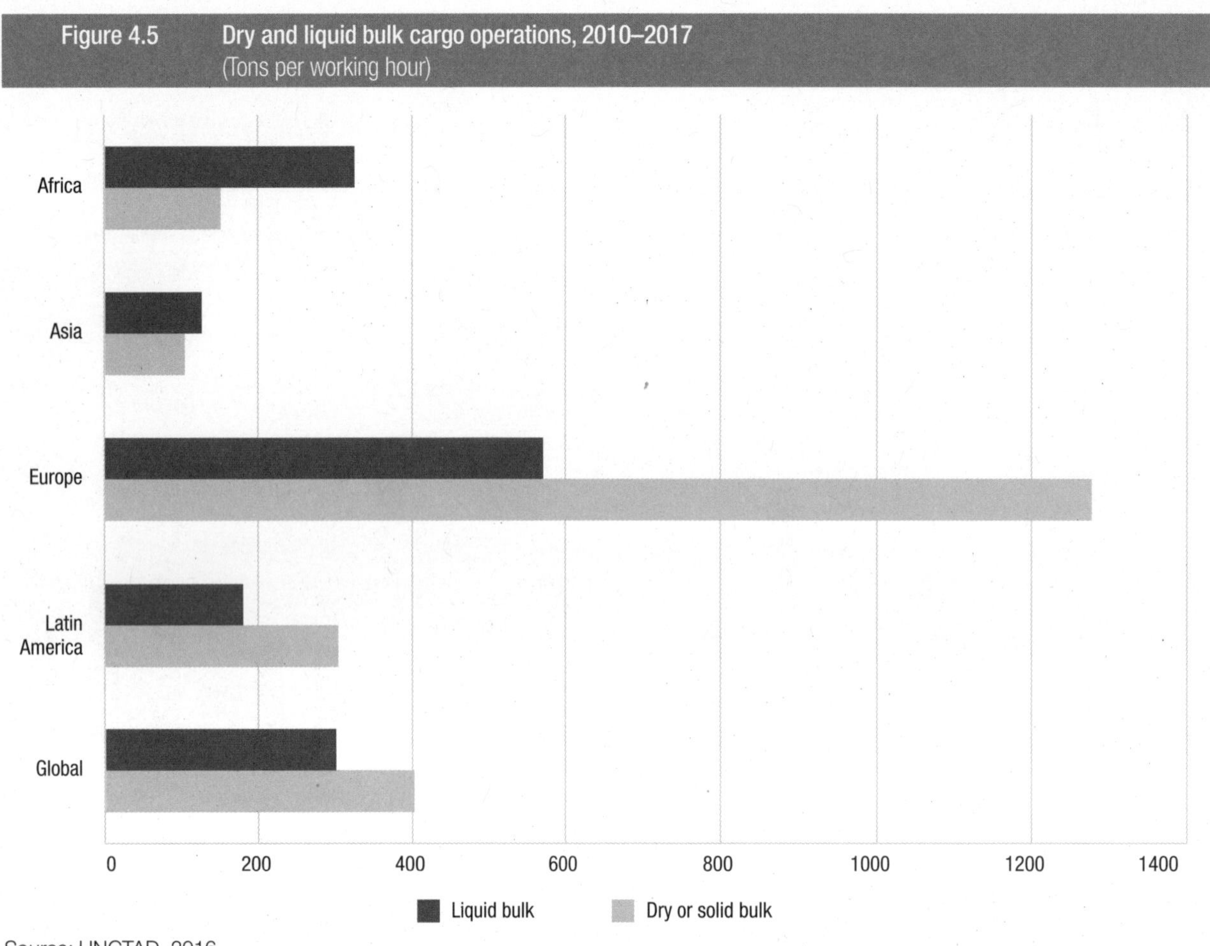

*Source:* UNCTAD, 2016.

Figure 4.6 Training costs as a percentage of wages, 2010–2017

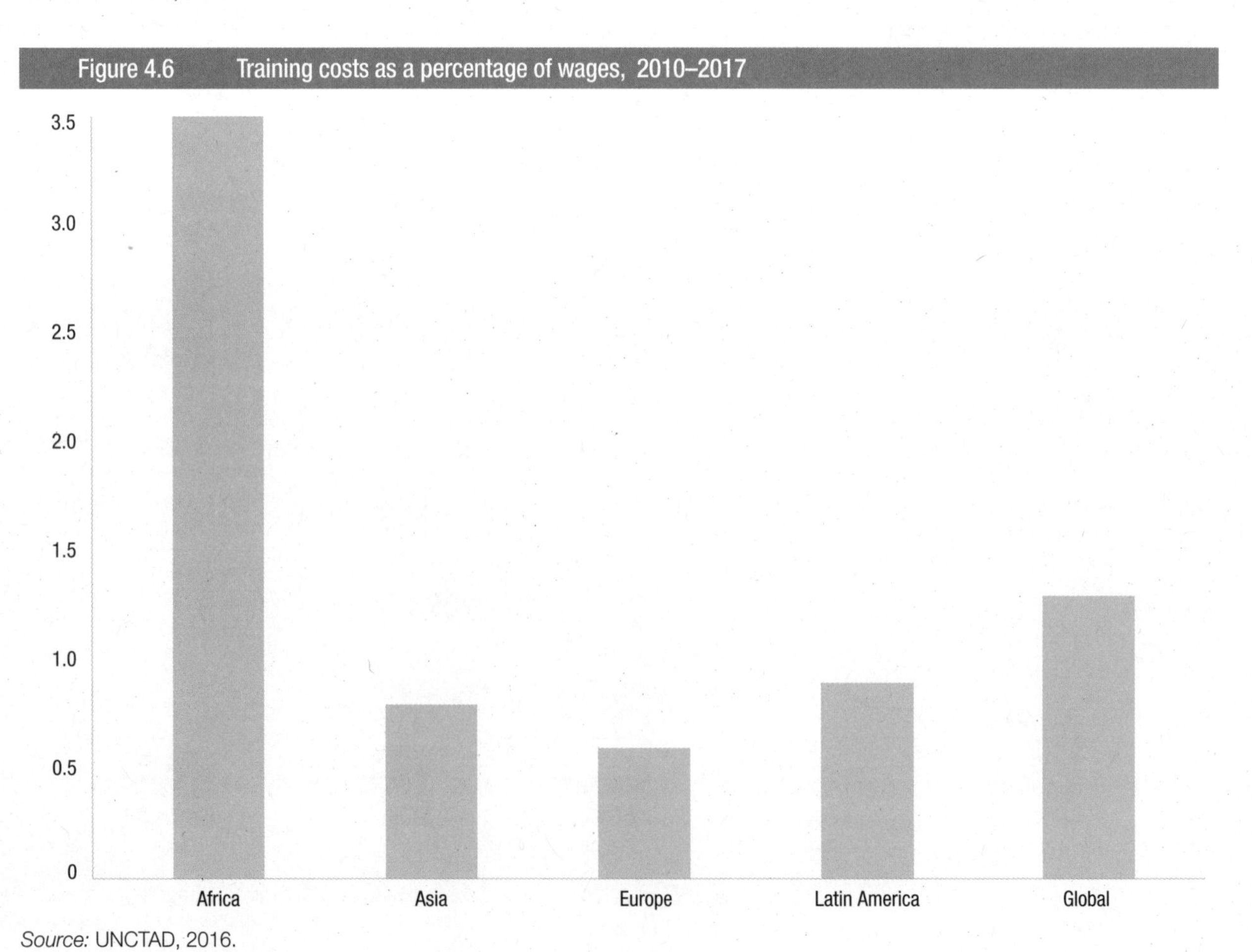


*Source:* UNCTAD, 2016.

**Box 4.1 UNCTAD port performance scorecard indicators**

Train for Trade is a component of the UNCTAD Port Management Programme, which supports port communities in developing countries seeking to ensure efficient and competitive port management, and in turn, support trade and economic development. The Programme creates port networks bringing together public, private and international entities. The aim is for port operators from public and private entities worldwide to share knowledge and expertise and to capitalize on research conducted in port management and port performance indicators (UNCTAD, 2016). For over 20 years, the Programme has provided training and capacity-building activities for four language networks (English, French, Portuguese and Spanish); 3,500 port managers from 49 countries in Africa, developing America, Asia, the Caribbean and Europe; and 110 replication cycles of one to two years at the national level. The Programme is recognized by beneficiaries, donors, partners and evaluators as a successful model of technical assistance. Under the activities of the Programme, UNCTAD has initiated work on port performance measurement. Starting in 2014, a series of international conferences brought together over 200 representatives from 30 member countries of the four language networks. The aim was to identify the port performance indicators that should be collected, the corresponding definitions, the underlying methodology and the technology to be adopted. The latter aims to ensure a common denominator across the various ports of the network of the Programme to promote meaningful comparisons.

One of the challenges faced by the Programme was the ability to discriminate results at the port level instead of country level. This is often the case with indicators such as the logistics performance index (World Bank), the global competitiveness index (World Economic Forum) and the liner shipping connectivity index (UNCTAD). These indicators are aggregated at the country level and do not provide a port-level perspective.

Additional information about the UNCTAD Port Management Programme and port performance scorecard is available at https://learn.unctad.org/course/index.php?categoryid=2.

*Source*: UNCTAD, 2017a.

## B. GLOBAL CONTAINER PORTS

Container port throughput is driven to a large extent by developments in the world economy and global demand, including investment, production and consumption requirements. Trans-shipment is a major area of container port activity that results in particular from hub-and-spoke container networks and could be enhanced by the further deployment of ultralarge container vessels. Trends in 2016 and 2017 point to the strategic importance of containerized port activity. Some 873 ports worldwide received regularly scheduled calls from fully cellular container ships across 141 countries, leading to over 560,000 individual port calls (Clarksons Research, 2017).

### 1. Increase in global container port throughput

UNCTAD estimates that global container port throughput rose by 6 per cent in 2017, three times the rate of 2016 (table 4.3). Increased port activity reflected the recovery of the world economy and the associated increase in trade flows. According to UNCTAD calculations, 752.2 million TEUs were handled by container ports in 2017. This total reflects the addition of some 42.3 million TEUs, an amount comparable to total container volumes handled by Shanghai, the top-ranked global port in volume terms.

Key factors contributing to higher volumes included strong growth on the intra-Asian trade route; improved consumer demand in the United States and Europe; and an increase in North–South trade volumes, which was supported by higher commodity export earnings in Africa and developing America, thus stimulating imports. However, the relatively rapid growth achieved by container ports after the weak performance of 2015 and 2106, suggests that apart from the cyclical recovery, some supply chain restocking may have further supported growth in 2017. Trans-shipment declined slightly from 26 per cent in 2016 to 25.8 per cent in 2017. While the configuration of capacity along shipping networks has reached a level of stability, the expansion of the Panama Canal could imply more direct calls to the East Coast of the United States and probably slower growth in trans-shipment activity in the Panama Canal and Caribbean region.

**Table 4.3 World container port throughput by region, 2016–2017**
(20-foot equivalent units and annual percentage change)

| | 2016 | 2017 | Annual percentage change |
|---|---|---|---|
| Asia | 454 513 516 | 484 176 997 | 6,5 |
| Africa | 30 406 398 | 32 078 811 | 5,5 |
| Europe | 111 973 904 | 119 384 254 | 6,6 |
| North America | 54 796 654 | 56 524 056 | 3,2 |
| Oceania | 11 596 923 | 11 659 835 | 0,5 |
| Developing America | 46 405 001 | 48 355 369 | 4,2 |
| **World total** | **709 692 396** | **752 179 321** | **6,0** |

*Source:* UNCTAD secretariat calculations, based on data collected by various sources, including Lloyd's List Intelligence, Jean-Paul Rodrigue, Hofstra University, Dynamar BV, Drewry Maritime Research and information posted on websites of port authorities and container port terminals.

*Note:* Data are reported in the format available. In some cases, country volumes were derived from secondary sources and reported growth rates. Country totals may conceal the fact that minor ports may not be included. Therefore, data in the table may differ from actual figures in some cases.

Asia plays a central role in global trade and shipping, as shown by activity in the container shipping sector. The Asia–Pacific region accounts for over 42 per cent of the number of ports and 60 per cent of the calls, with China representing 19 per cent of all calls alone (Clarksons Research, 2017). These trends have been largely supported by globalization. The second most important player is Europe, which accounts for 28 per cent of world container ports and 21 per cent of port calls.

In line with trends in port calls, Asia dominates the container-handling business. The region continued to account for nearly two thirds of the global container port throughput (figure 4.7). Volumes handled in the region increased by 6.5 per cent. Some 240 million TEUs were recorded in China, including Hong Kong, China and Taiwan Province of China. This represents almost half of all port volumes handled in the region. Restrictions imposed by the Government of China limiting imports of some waste material on the backhaul journeys from North America and Europe are likely to increase the incidence of empties in the overall traffic handled by ports, which could exacerbate the trade and freight rate imbalances on the trans-Pacific route.

Elsewhere in Asia, container port throughput in 2017 was influenced by developments in the Islamic Republic of Iran and sanctions imposed on Qatar. While volumes in Bandar Abbas port increased by over 20 per cent, the imposition of sanctions on the Islamic Republic of Iran had already started to weigh on port performance in late 2017 (Drewry Maritime Research, 2018a). Jebel Ali faced some competition from Bandar Abbas port, despite increasing volumes by 4 per cent over 2016. Port Sohar in Oman gained the most from sanctions imposed on Qatar. Growth in South Asia surpassed 10.7 per cent, reflecting among other factors, the growing shift of manufacturing towards Bangladesh, India and Pakistan. In India, Jawaharlal Nehru Port terminals attracted 4.8 per cent more business in 2017. A new container terminal in Jawaharlal Nehru Port, which has been running close to design capacity for several years, was opened in early 2018.

**Figure 4.7 World container port throughput by region, 2017**
(Percentage share in total 20-foot equivalent units)

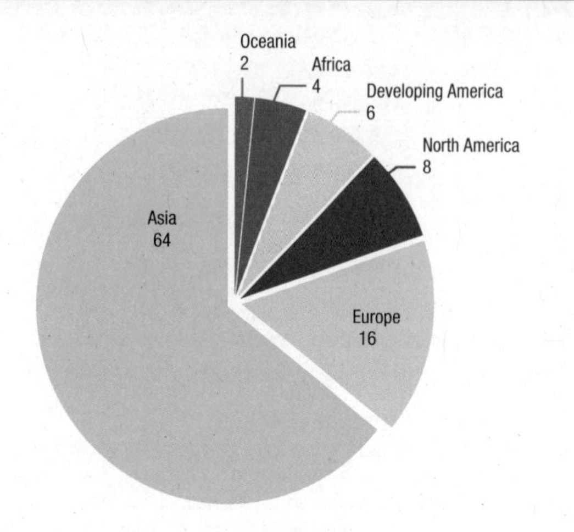


*Source:* UNCTAD secretariat calculations, derived from table 4.3.

Reflecting to a large extent the recovery in the European Union in 2017, volumes handled in European ports increased by 6.6 per cent. With volumes reaching nearly 120 million TEUs, Europe accounted for 16 per cent of global container port throughput.

A development affecting European ports during the year was the growing presence of the China Ocean Shipping Company as a principal port investor. After acquiring port facilities in Greece, Italy and Spain, the company established a presence in Northern Europe by signing a concession agreement with Zeebrugge Port Authority to open a container terminal – this was made possible in part by the Belt and Road Initiative. The company is expected to emerge as a world leader among terminal operators by 2020 (Wei, 2018).

North America maintained an 8 per cent share of total container port volumes, supported by strong activity in the United States. Africa's share of world container port throughput was estimated at 4 per cent, surpassing Oceania's 2 per cent share. However, this was still below the 6 per cent accounted for by developing American ports. Volumes in Africa increased due to stronger import demand. Many sub-Saharan African countries experienced a higher demand for their exports and recorded better export earnings than in the past. This in turn boosted imports, with the southbound Asia–West Africa trade growing at its fastest rate since 2014 (Drewry Maritime Research, 2017a). This is reflected in increased throughputs in South Africa and Western Africa, in contrast with losses incurred in 2016. In particular, the recovery in Angola and Nigeria from a low-price environment and the robust economies of Côte d'Ivoire and Ghana contributed favourably to a 9.5 per cent increase in West African ports' container throughput.

In Australia and New Zealand, growth in container port volumes was sustained by external demand and strong consumer spending, while in developing America, volumes were driven by the higher commodity prices environment and the end of recession in key economies such as in Brazil. Container traffic from Asia to the East Coast of South America bounced back in 2017, expanding by 15.5 per cent. The recovery was driven by Brazilian imports, which rose sharply, by 22 per cent.

As shown in table 4.4, container port activity tends to be concentrated in major ports. These are generally mega ports, which serve as hubs or gateways for important hinterlands (Clarksons Research, 2017). The combined throughput at the world's leading 20 container terminals increased by 5.9 per cent. Together, they handled an estimated 336.6 million TEUs, accounting for 45 per cent of the world's total. Except for the ports of Klang and Kaohsiung, all ports in the ranking recorded volume gains. The contribution of Asian container ports

| Table 4.4 | Leading 20 global container ports, 2017 (Thousand 20-foot equivalent units, percentage annual change and rank) | | | | |
|---|---|---|---|---|---|
| **Port** | **Economy** | **Throughput 2017** | **Throughput 2016** | **Percentage change 2016–2017** | **Rank 2017** |
| Shanghai | China | 40 230 | 37 133 | 8,3 | 1 |
| Singapore | Singapore | 33 670 | 30 904 | 9,0 | 2 |
| Shenzhen | China | 25 210 | 23 979 | 5,1 | 3 |
| Ningbo-Zhoushan | China | 24 610 | 21 560 | 14,1 | 4 |
| Busan | Republic of Korea | 21 400 | 19 850 | 7,8 | 5 |
| Hong Kong | Hong Kong SAR | 20 760 | 19 813 | 4,8 | 6 |
| Guangzhou (Nansha) | China | 20 370 | 18 858 | 8,0 | 7 |
| Qingdao | China | 18 260 | 18 010 | 1,4 | 8 |
| Dubai | United Arab Emirates | 15 440 | 14 772 | 4,5 | 9 |
| Tianjin | China | 15 210 | 14 490 | 5,0 | 10 |
| Rotterdam | Netherlands | 13 600 | 12 385 | 9,8 | 11 |
| Port Klang | Malaysia | 12 060 | 13 170 | -8,4 | 12 |
| Antwerp | Belgium | 10 450 | 10 037 | 4,1 | 13 |
| Xiamen | China | 10 380 | 9 614 | 8,0 | 14 |
| Kaohsiung | Taiwan Province of China | 10 240 | 10 465 | -2,2 | 15 |
| Dalian | China | 9 710 | 9 614 | 1,0 | 16 |
| Los Angeles | United States | 9 340 | 8 857 | 5,5 | 17 |
| Hamburg | Germany | 9 600 | 8 910 | 7,7 | 18 |
| Tanjung Pelepas | Malaysia | 8 330 | 8 281 | 0,6 | 19 |
| Laem Chabang | Thailand | 7 760 | 7 227 | 7,4 | 20 |
| Total | | 336 630 | 317 929 | 5,9 | |

*Source:* UNCTAD secretariat calculations, based on various industry sources.
*Abbreviation:* SAR, Special Administrative Region.

surpasses all other regions, as 80 per cent of the ports featuring in the top 20 are Asian. Nearly two thirds of these are in China.

Apart from the contraction in volumes experienced by the ports of Klang and Kaohsiung, growth of individual ports varied between a low of 0.6 per cent in Tanjung Pelepas and 14.1 per cent in Ningbo-Zhoushan. Shanghai remained the busiest container port worldwide; volumes handled expanded by 8.3 per cent, bringing the total volume to 40.2 million TEUs. Singapore ranked second, handling 33.7 million TEUs, a 9 per cent increase over 2016. In third position, the amount of volumes handled by Shenzhen increased by 5.1 per cent, to 25.2 million TEUs. Ranked fourth, Ningbo-Zhoushan saw the largest increase in volumes, which rose by 14.1 per cent to 24.6 million TEUs. As the biggest receiver of plastic waste, Guangzhou, and to some extent, Shenzhen, which imports wastepaper, are likely to be affected by a new regulation introduced in China in late 2017, limiting the imports of some types of wastes (Drewry Maritime Research, 2017a). Outside Asia, four ports, Rotterdam, Antwerp, Los Angeles and Hamburg, are among the top 20 ports. All four handled larger volumes in 2017, although Rotterdam saw the largest increase, as cargo throughput expanded by nearly 10 per cent, above levels in 2016.

## 2. Operational performance of world container ports

Strategic liner shipping alliances and the associated trend of vessel upsizing have added complexity to the container shipping and port relationship and triggered new dynamics where shipping lines have greater bargaining power and influence.

Vessel size increases and the rise of mega alliances have heightened the requirements for ports to adapt and respond to more stringent requirements. Bigger call sizes exert additional pressure on ports and terminals and require an effective response measure to ensure that space, equipment, labour, technology and port services are optimized. This raises the question of whether costs and benefits associated with the upsizing of vessels and alliances are fairly distributed between shipping lines and ports.

Liner shipping consolidation, alliance formation and the deployment of larger vessels have combined, leading to greater competition among container ports to win port calls (Notteboom et al., 2017). For example, the port of Klang handled less cargo during the year, as alliance members limited their port calls. Meanwhile, the ports of Singapore and Tanjung Pelepas recorded an increase of 8.2 per cent and 3.4 per cent, respectively, following the decision by shipping alliance members to use them as pivotal ports of call (Shanghai International Shipping Institute, 2017).

As ports compete for fewer services by larger vessels, ports and terminals are interacting with carriers that have strong negotiating and decision-making power. The stakes are high for terminal operators, as a call made by alliance members using larger vessels can generate significant port volumes and business. For example, a weekly call concerning one of the services between Northern Europe and the Far East is estimated to result in annual container volumes of about 300,000 TEUs per port of call. A liner service using ships with only a capacity of 20,000 TEUs could increase this estimate to an average of about 450,000 TEUs per year per port of call (Notteboom et al., 2017).

The dynamics between shipping lines and container port terminals is further shaped by the ability of lines to take part in port operations though shareholdings and joint ventures with terminal operators, sister companies or subsidiaries involved in terminal operations. This can affect approaches to terminal concessions. Although a terminal operator owned by a shipping line may have a more stable cargo base, regulators may prefer that concessions be granted to independent operators to allow access to all port-handling service providers.

Some of these concerns, including the operational challenges arising from the growing use of mega ships and formation of mega alliances, are reflected in port productivity and performance patterns. While liner shipping networks seem to have benefited from efficiencies arising from consolidation and alliance restructuring, gains at the port level have not evolved at the same pace. Container berth productivity is constrained by the growing volume of boxes exchanged in vessel calls during peak hours (Fairplay, 2018). The deployment of larger vessels and alliance network design have direct implications for the number of boxes exchanged per call, which in turn, exerts additional pressure on ports' handling capacities.

Existing data for 2017 indicate an annual global increase of 9 per cent in the number of containers handled per call. Northern European ports experienced the largest growth – 20 per cent – in average call sizes, compared with 2016. In comparison, call sizes at ports in South-East Asia and developing America increased by 11 per cent in each region. Elsewhere, results were less positive, showing no growth (Africa) or modest declines (Oceania). With regard to results in individual container ports and terminals, the largest increases in call sizes were seen in Antwerp (29 per cent), Yangshan (27 per cent) and Manila (22 per cent) (Fairplay, 2018).

The need to handle more containers at the same time exerts pressure on berth and yard operations. While the increased demand for cargo-handling operations can be mitigated to some extent through the container distribution in ship-planning processes, larger call sizes, combined with a limited number of cranes, reduces optimal crane intensity. The gap between growth in call size and productivity widens when the number of boxes exchanged exceeds 4,000 (Fairplay, 2017b). Some observers contend that ports perform best when ship sizes are within the range of 4,000–14,000 TEUs. These sizes are optimal for quayside performance, although they allow for fewer rows of containers than larger ships. Performance of ships with a capacity of more than 14,000 TEUs is negatively affected by the pressure on equipment and space, for example spreaders, trolley distances, berth and yard areas.

Global port productivity fell in 2017, indicating that container terminals were challenged by the deployment of larger vessels and the growth in port call sizes. In this context, port productivity refers to the number of container moves per hour of time spent by vessels in port, weighted by the call size, which is significantly impacted by the number of cranes deployed to service a ship. Bearing these considerations in mind, some estimates for 2017 indicate a 3 per cent average drop in weighted port productivity globally, compared with 2016 (JOC.com, 2018).

The decline in port productivity affected all regions. One of the steepest declines was experienced in Africa, where port productivity dipped by 12 per cent. Productivity fell by more than 7 per cent in developing America, Western Asia and Indian ports. The impact on European and North American ports was less pronounced, with reductions of 3 per cent in the number of container moves per hour spent by vessels in time at berth. South-East Asia was the only region where some port productivity gains were achieved, despite an increase in call sizes. In terms of individual ports, the greatest declines in port productivity were seen in Manila (21 per cent), and in Dalian and Laem Chabang, where productivity declined by 16 per cent. On the other hand, some ports such as Long Beach, California and Chiwan, China recorded an increase in productivity.

Interestingly, both the number of moves per total hours spent by vessels in port and the waiting time between arrival and the allocation of berth decreased, the latter by 6 per cent worldwide (JOC.com, 2018). The world's largest ports recorded a reduction in the port-to-berth time; the largest improvements were witnessed in the ports of Antwerp and Hamburg. Less positive performances were recorded elsewhere. For example, berth-waiting times more than doubled in Manila and increased almost by half in the port of Shekou. Increases in port-to-berth waiting times were also recorded in India and some African countries.

The performance of major trans-shipment hubs was reported to be relatively even among the various ports. The average port-to-berth waiting time in Jebel Ali was estimated at 2.7 hours, while in Hong Kong (China), Busan and Singapore, waiting times averaged about 2.4 hours. The competitiveness of ports such as Tanjung Pelepas and Klang could be observed with waiting times of 2.2 hours and 2.4 hours, respectively. The average waiting time at Tanjung Priok, which attracted mainline calls in 2017, was also 2.4 hours.

Table 4.5 shows the average time in port by vessel type at the global level. In 2017, the average time in port for all ships was estimated at 31.2 hours, an improvement over the previous year when ships stayed an average of 33.6 hours in ports. Containerized vessels tend to spend less time in ports, followed by dry cargo ships, gas carriers and tankers. Bulk carriers experience the longest time in port, about 65 hours on average, more than double the global average for all ships.

| Table 4.5 | Average time in port, world, 2016 and 2017 | | | |
|---|---|---|---|---|
| | Days in port | | Total arrivals | Total deadweight tonnage (thousands of tons) |
| Vessel type | 2016 | 2017 | 2017 | 2017 |
| Container ships | 0,87 | 0,92 | 447 626 | 18 894 342 |
| Tankers | 1,36 | 1,30 | 301 713 | 9 648 282 |
| Gas carriers | 1,05 | 1,10 | 64 603 | 890 880 |
| Bulk carriers | 2,72 | 2,68 | 236 407 | 13 152 509 |
| Dry cargo and passenger ships | 1,10 | 1,02 | 3 995 242 | 7 280 933 |
| Total | 1,37 | 1,31 | 5 045 591 | 49 866 946 |

*Source:* Data provided by Marine Traffic, 2018.

*Notes:* Averages refer to medians. Time in port is defined as the difference between the time that the ship enters the port limits (excluding anchorages) and the time that the ship exits those limits. Irrespective of whether the ship's visit is related to cargo operations or other types of operations such as bunkering, repair, maintenance, storage and idling, time in port includes the time prior to berthing, the time spent at berth (dwell and working times) and the time spent undocking and transiting out of port limits.

Aside from typical operational and service level indicators, such as crane moves per hour and berth allocation waiting time, port performance can also be assessed according to the intensity of port asset utilization. Quay lines, cranes and land are important and expensive assets, for which the level of utilization is a key performance indicator, especially for investors. As gantry crane expenditure hovers around $10 million per crane and quay construction can cost as much as $100,000 per metre – the greater the utilization levels, the higher the performance of these assets (Drewry Maritime Research, 2017b).

Table 4.6 features relevant industry benchmarks and design parameters generally used to measure intensity usage of assets and performance. Table 4.7 reviews the asset use intensity between 2013 and 2016. It shows that asset use intensity remained unchanged overall, although land use intensity decreased. On a global basis, the intensity of quay line usage typically achieved by terminals worldwide is estimated at 1,100 TEUs per metre per year. As shown in table 4.6, the actual performance in 2016 was about 1,150 TEUs per metre, an intensity usage below the theoretical design parameter of 1,500 TEUs per metre. That said, performance varied at some terminals, especially in Asia, where it was relatively better than typical industry performance. Quay line performance above 2,000 TEUs per metre per year were observed in the ports of Busan; Singapore; Shanghai; Ningbo-Zhoushan; Hong Kong, China; Klang; Laerm Chabang; and Jawaharlal Nehru Port Terminal. Many of these also reached more than 250,000 TEUs per crane per year, and more than 50,000 TEUs per hectare per year (Drewry Maritime Research, 2017b).

Overall, the deployment of larger container ships in recent years seems to have had little impact on the annual use of quay line assets and on TEUs handled per gantry crane, whose levels generally stood at some 127,000 TEUs per crane a year. Land use intensity declined slightly, averaging close to 27,000 TEUs per hectare per year in 2016. This may reflect the impact of the growing size of ships calling at ports and the associated pressure on yard operations during periods of peak volumes.

An increase in yard space to alleviate pressure can have the effect of reducing intensity usage. However, other factors may also affect land usage, as shown in North America, where a shift from chassis operations towards fully rounded yard systems improved port performance (Drewry Maritime Research, 2017b). Similarly, ports in developing America improved land usage by increasingly moving away from small multi-purpose terminals in many locations towards larger, specialized container terminals. A terminal's size can also influence usage performance, as illustrated by the relatively higher performance observed in Asia. A terminal's function also has a role to play, with trans-shipment ports generally performing at higher levels than gateway ports. Operational factors such as cargo-handling equipment and working hours tend to have a strong impact on asset usage indicators such as TEUs handled per hectare, per metre of quay line and per crane.

| Table 4.6 | Usage intensity of world container terminal assets, 2016 | | |
|---|---|---|---|
| Measure per annum | Typical industry design parameters | Performance | Remarks |
| TEUs per metre of quay | 1 500 | 1 154 | Design parameters typically range from 800–1700 TEUs per metre per year |
| TEUs per ship to shore gantry crane | 200 000 | 127 167 | Design parameters are influenced by ratio of number of boxes to TEUs |
| TEUs per hectare | 40 000 | 26 366 | Design parameters are highly dependent on yard equipment type and dwell times |

*Source:* Drewry Maritime Research, 2017b.

*Note:* Figures on actual performance are based on a sample of 321 terminals handling over 200,000 TEUs per annum.

Table 4.7 Usage intensity of world container terminal assets by region, 2003 and 2016

| Region | 2003 | 2016 | Percentage change |
|---|---|---|---|
| **Developing America** | | | |
| TEUs per metre of quay per annum | 665 | 849 | 27,7 |
| TEUs per ship to shore gantry crane per annum | 105 517 | 110 307 | 4,53 |
| TEUs per hectare per annum | 16 696 | 27 752 | 66,2 |
| **Europe** | | | |
| TEUs per metre of quay per annum | 653 | 761 | 16,53 |
| TEUs per ship to shore gantry crane per annum | 100 110 | 94 819 | -5,28 |
| TEUs per hectare per annum | 16 651 | 18 794 | 12,87 |
| **North America** | | | |
| TEUs per metre of quay per annum | 665 | 777 | 16,8 |
| TEUs per ship to shore gantry crane per annum | 90 661 | 91 885 | 1,4 |
| TEUs per hectare per annum | 9 604 | 14 407 | 50,0 |

*Source:* Drewry Maritime Research, 2017b.
*Note:* Figures on actual performance are based on a sample of 321 terminals handling over 200,000 TEUs per annum.

## C. GLOBAL DRY BULK TERMINALS

### 1. Global dry bulk terminals benefit from growing demand for raw materials and energy

Positive trends in population growth, urbanization, infrastructure development, construction activity, and industrial and steel output, especially in rapidly emerging developing countries in Asia, have generally had a marked impact on bulk terminals worldwide. Dry bulk commodities have been the mainstay of international seaborne trade volumes in recent years, accounting for almost half of world seaborne trade flows in 2017.

Trends in coal trade volumes in 2017 were shaped by growing environmental sustainability imperatives. Many countries continued their energy transition towards less carbon-intensive, cleaner sources of energy, thereby lessening the demand for coal. While this may be true in terms of coal imports received in Europe, coal remained a major source of energy in many developing countries and a key export commodity for countries such as Australia, Colombia and Indonesia. For countries in South-East Asia, notably Indonesia, the Republic of Korea and Viet Nam, coal remained a key cargo import.

China remained the leading source of global import demand for iron ore, (see chapter 1). With regard to exports, Australia and Brazil remained the main players. Table 4.8 features some major dry bulk terminals and highlights the central role of countries such as Australia, China, Indonesia, the Russian Federation and the United States, as well as Northern European countries as main loading and unloading areas for major dry bulk commodities.

Dry bulk throughput at major world ports showed divergent growth. Throughput at Qinhuangdao, reflecting China's importance as the main market for iron ore, grew by 46 per cent between 2016 and 2017. Dry bulk throughput at major ports in Australia, notably at Port Hedland – the country's largest export facility and the world's largest iron ore loading terminal (Business Insider Australia, 2017) – continued to increase with an annual growth rate of 5.5 per cent. Three major global mining companies (Broken Hill Proprietary Billiton, Hancock Prospecting and Fortescue Metals Group) are using the port. Rio Tinto, however, is using another port (Port Dampier) (Market Realist, 2018). In Singapore, growth in volumes remained stable. While overall cargo volumes handled have grown steadily over the past few years, the port is said to be increasingly focused on trade in liquefied natural gas (Fairplay, 2017a). Rotterdam, the biggest and busiest port in Europe, recorded a slight decrease in throughput, reflecting reduced demand for European coal imports.

### 2. Performance of selected global dry bulk terminals

Being able to monitor and assess the performance of bulk terminals, including dry bulk terminals, is important for planning, investment, safety, productivity and service quality. To this end, the Baltic and International Maritime Council (BIMCO) launched a vetting system of dry bulk terminals around the world in 2015 (BIMCO, 2017). Relying upon reports by shipowners about their ships' visits to dry bulk terminals at the global level, the vetting scheme is considered useful in gathering information about terminal performance and highlighting areas that require further monitoring and improvement. Data collected between 2015 and 2017 focused on parameters such as mooring and berth arrangements, terminal services, equipment, information exchanges

**Table 4.8 Main dry bulk terminals: Estimated country market share in world exports by commodity, 2017**
(Percentage)

| Iron ore | Percentage | Coal | Percentage | Grain | Percentage |
|---|---|---|---|---|---|
| **Australia** | 56,2 | **Australia** | 30,3 | **United States** | 27,7 |
| Cape Lambert | | Abbott Point | | Corpus Christi | |
| Dampier | | Dalrymple Bay | | Galveston | |
| Port Hedland | | Gladstone | | Hampton Roads | |
| Port Latta | | Hay Point | | Houston | |
| Port Walcott | | Newcastle | | New Orleans | |
| Yampi Sound | | Port Kembla | | Norfolk | |
| | | | | Portland | |
| **Brazil** | 25,8 | **Indonesia** | 30,4 | | |
| Ponta da Madeira | | Balikpapan | | **European Union** | 9,8 |
| Ponta do Ubu | | Banjamarsin | | Immingham | |
| Sepetiba | | Kota Baru | | Le Havre | |
| Tubarao | | Pulau Laut | | Muuga | |
| | | Tanjung Bara | | Rouen | |
| **South Africa** | 4,4 | Tarahan | | Klaipeda | |
| Saldanha Bay | | | | Riga | |
| **Canada** | 2,8 | **Russian Federation** | 11,4 | **Argentina** | 10,9 |
| Port Cartier | | Vostochny | | Bahia Blanca | |
| Seven Islands | | Murmansk | | Buenos Aires | |
| | | | | La Plata | |
| **Ukraine** | 0,7 | **Colombia** | 7,1 | Necochea | |
| Yuzhny | | Cartagena | | Parana | |
| Illichevsk | | Puerto Bolivar | | Rosario | |
| | | Puerto Prodeco | | | |
| **Sweden** | 1,5 | Santa Marta | | **Australia** | 9,1 |
| Lulea | | | | Brisbane | |
| Oxelsund | | **South Africa** | 6,8 | Geraldton | |
| | | Durban | | Melbourne | |
| **Chile** | 1,0 | Richards Bay | | Port Giles | |
| Caldera | | | | Port Lincoln | |
| Calderilla | | **United States**[a] | 6,9 | Sydney | |
| Chanaral | | Baltimore | | Wallaroo | |
| | | Corpus Christi | | | |
| **Iran (Islamic Republic of)** | 1,3 | Long Beach | | **Canada** | 7,0 |
| Bandar Abbas | | Los Angeles | | Halifax | |
| | | Mississippi River System terminals | | Baie Comeau | |
| **Mauritania** | 0,8 | Mobile | | Prince Rupert | |
| Nouadhibou | | Newport News | | Vancouver | |
| | | Norfolk | | | |
| **Peru** | 1,0 | Seward | | **Russian Federation** | 10,2 |
| San Nicolas | | Stockton | | Novorossiysk | |
| | | | | Rostov | |
| | | **Canada**[b] | 2,3 | | |
| | | Canso Anchorage | | | |
| **India** | 2,0 | Neptune Terminal | | **Ukraine** | 12,6 |
| Mormogao | | Prince Rupert | | Odessa | |
| Calcutta | | Roberts Bank | | Nikolaev | |
| Paradip | | | | Ilychevsk | |
| New Mangalore | | **China** | 0,3 | | |
| Chenai | | Dalian | | | |
| Kakinada | | Qingdao | | | |
| | | Qinhuangdao | | | |
| | | Rizhao | | | |
| | | **Mozambique** | 0,4 | | |
| | | Maputo | | | |
| | | Beira | | | |

*Source:* UNCTAD secretariat calculations, based on data from Clarksons Research, 2018.

[a] *Excluding* exports to Canada.

[b] *Excluding* exports to the United States.

between ships and terminals, and loading and unloading cargo handling. By 1 December 2017, 27 ports had more than five entries or reports. None of the ports had ratings below average. Scores were based on a weighting system where loading and unloading had the highest value, followed by mooring and berth arrangements, and information exchanges.

The three leading dry bulk terminals according to the BIMCO vetting scheme were Santander and Bilbao, Spain and Quebec, Canada. Santander ranked first in terms of terminal handling of loading and unloading operations, terminal mooring and berthing arrangements, and information exchanges between ships and terminals, and terminal equipment. According to the 2017 vetting report, over 93 per cent of ports in the analysis received an average score or better in terms of communications between ships and terminals, loading and unloading activity, and standards and maintenance of equipment. Areas requiring further improvement relate to challenges arising from the need for language skills, permanent pressure on ship crews and masters, unexpected claims, and unnecessary bureaucratic and aggressive port authorities (BIMCO, 2017). In addition, ports rated poorly when the cost of terminal services was either too high or the service was non-existent. While the vetting report is useful, there are limitations to the system. Additional data and reports would be required to improve the statistical validity and reliability of results obtained.

## D. DIGITALIZATION IN PORTS

A factor that is evolving at an accelerated pace with potentially profound implications for port operations and management is digitalization. There is no widely accepted definition of the digital economy. The latest developments in digitalization are emerging from a combination of technologies that are becoming more pervasive across mechanical systems, communications and infrastructure (UNCTAD, 2017b). Key technologies supporting digitalization in maritime transport include innovations such as the Internet of things, robotics, automation, artificial intelligence, unmanned vehicles and equipment, and blockchain (see chapters 1, 2 and 5).

The application of such innovations in ports permeates all aspects of a port business, including operations, planning, design infrastructure development and maintenance. They bring new opportunities for ports by unlocking more value that extends beyond traditional cargo-handling activities. Relevant technologies can help optimize traffic; increase operational efficiency, process transparency and speed; automate processes; and reduce inefficiencies and errors. Concrete examples of ways in which the impact of innovative technologies will likely be felt in ports include changes to loading and unloading operations (machine-to-machine communication, platform solutions, robotics, intelligent asset development and mobile workforces), storage (big data analytics, smart metering and single views of stock) and industrial processing (smart grids, smart energy management, three-dimensional printing, safety analytics and predictive maintenance).

The maritime transport industry is increasingly playing catch-up when it comes to enhancing the use of innovative technologies to improve systems and processes. One industry survey reveals that according to 15 per cent of respondents, autonomous terminal equipment was already being used (Vonck, 2017). According to 9 per cent of the respondents, autonomous drones for port services are being used, while 43 per cent consider this a short-term trend. Respondents generally agreed that irrespective of the speed at which digitalization unfolds, there is a growing need to upgrade skills and enhance expertise, efficiency and knowledge.

A review of ports around the world indicates that the sector has embraced technology to a certain extent, with operations of many ports having changed dramatically over the past few decades. For example, scanning technologies are increasingly being used for security and trade facilitation, while automation is being introduced in various container terminals. A focus on container port terminals around the world provides a good overview of the actual state of play. Container terminal automation – the use of robotized and remotely controlled handling systems along with the transition from manual to automated processes – is still at relatively early stages of utilization, as 97 per cent of world container port terminals are not automated. The share of container terminals that are fully automated is estimated at 1 per cent, while semi-automated terminals account for 2 per cent thereof (Drewry Maritime Research, 2018b). Table 4.9 provides an overview of the main terminals where full or partial automation is being implemented or planned. Fully automated terminals are those where the yard stacking and the horizontal transfer between the quay and the yard is automated, while semi-automated terminals are those where only the yard stacking is automated.

Container terminals are increasingly using higher levels of automation to improve productivity and efficiency and secure a competitive advantage. An industry survey reveals that nearly 75 per cent of terminal operators consider automation critical in order to remain competitive in the next three to five years, while 65 per cent view automation as an operational safety lever (Hellenic Shipping News, 2018). Over 60 per cent of respondent terminal operators expect automation to help improve operational control and consistency, while 58 per cent expect it to cut overall terminal operational costs. Respondents were positive about the potential return on investment overall. About one third of respondents see in automation a way to increase productivity by up to 50 per cent, while about one fifth believe that automation could reduce operational costs by more than 50 per cent.

**Table 4.9 Overview of automation trends in ports, 2017**

| Port | Terminal | Operational level of automation [a] |
|---|---|---|
| Brisbane, Australia | Container terminals, Fisherman Island Container Terminal | Semi |
| | Fisherman Island berths 8–10 | Fully |
| Melbourne, Australia | Victoria International Container Terminal | Fully |
| Sydney, Australia | Sydney International Container Terminals | Semi |
| | Brotherson Dock North | Fully |
| Antwerp, Belgium | Gateway | Semi |
| Qingdao, China | New Qianwan | Fully |
| Shanghai, China | Yangshan, phase 4 | Fully (trial vessels handled end-2017) |
| Tianjin, China | Dong Jiang | Not confirmed; in development |
| Xiamen, China | Ocean Gate Container Terminal [b] | Fully (phase 1 operational; phases 2 and 3 in development) |
| Hamburg, Germany | Altenwerder Container Terminal | Fully |
| | Burchardkai | Semi |
| Vizhinjam, India | Adani | Not confirmed; in development |
| Surabaya, Indonesia | Lamong Bay and Petikemas | Semi |
| Dublin, Ireland | Ferryport Terminals | Semi; planned |
| Vado Ligure, Italy | APM Terminals | Semi; due to be operational 2018 |
| Nagoya, Japan | Tobishima Pier South Side Container Terminal | Fully |
| Tokyo, Japan | Oi Terminal 5 | Semi |
| Lázaro Cárdenas, Mexico | Terminal 2 | Semi |
| Tuxpan, Mexico | Port Terminal | Semi |
| Tanger Med, Morocco | Tanger Med 2 | Not confirmed; due to open 2019 |
| Rotterdam, Netherlands | "Delta Dedicated East and West Terminals, Euromax, World Gateway and APM Terminals" | Fully |
| Auckland, New Zealand | Fergusson Container Terminal | Semi; due to be completed 2019 |
| Colón, Panama | Manzanillo International Terminal | Semi |
| Singapore | Pasir Panjang Terminals 1, 2, 3 and 4 | Semi |
| | Tuas | Not confirmed; planned |
| Busan, Republic of Korea | "Pusan Newport International and container terminal, Newport Company, Hanjin Newport Company and Hyundai Pusan Newport". | Semi |
| Incheon, Republic of Korea | Hanjin Incheon Container Terminal | Semi |
| Algeciras, Spain | Total Terminal Internacional | Semi |
| Barcelona, Spain | Europe South | Semi |
| Dubai, United Arab Emirates | Jebel Ali Terminals 3 and 4 | Semi (terminal 3 operational; terminal 4 due to be operational 2018) |
| Abu Dhabi, United Arab Emirates | Khalifa Container Terminal | Semi |
| Liverpool, United Kingdom | Liverpool 2 Container Terminal | Semi |
| London, United Kingdom | Dubai Ports London Gateway Container Terminal and Thamesport | Semi |
| Long Beach, United States | Container Terminal | Fully (Middle Harbour Redevelopment Project in development) |
| Los Angeles, United States | TraPac | Fully |
| New York, United States | Global Container Terminals | Semi |
| Norfolk, United States | Virginia International Gateway | Semi |
| | International Terminals | Semi; in development |
| Kaohsiung, Taiwan Province of China | Terminals 4 and 5 and Kao Ming Container Terminal | Semi |
| Taipei, Taiwan Province of China | Container Terminal | Semi |

*Source:* Drewry Maritime Research, 2018b.

[a] *Those* not yet fully operational are indicated.

[b] Also known as Yuanhai Automated Container Terminal. Double trolley quay cranes will have significant automation.

However, the advantages of automation in ports should be considered within context. In some cases, there can be a delay in reaching expected productivity levels due to many different innovations coming together without sufficient integration, and a lack of overall controllability. While technology is a key enabler, it is not the only parameter influencing terminal productivity (Linked in, 2018).

Reported challenges to wider implementation of port automation solutions include costs, shortage of skills or resources to implement and manage automation, concerns of labour unions and time required for implementation. With respect to labour, one study focusing on the maritime cluster in the Netherlands finds that the number of jobs in the maritime cluster will decrease by at least 25 per cent with the advent of automation. Jobs in the port sector are projected to drop by 8.2 per cent. By comparison, the number of jobs in shipping is expected to fall by 1.8 per cent. The analysis concludes that the largest subsectors at risk are ports, maritime suppliers and inland navigation (Vonck, 2017).

In sum, a broad range of technologies with applications in ports and terminals offers an opportunity for port stakeholders to innovate and generate additional value in the form of greater efficiency, enhanced productivity, greater safety and heightened environmental protection. For ports to effectively reap the benefits of digitalization, various concerns will need to be monitored and addressed. These include the potential regionalization of production and trade patterns associated with robotics and three-dimensional printing, potential labour market disruptions, regulatory changes and the need for common standards, in particular when applying blockchain technology and data analytics. To do so, it is essential to improve understanding of issues at stake, and strengthen partnerships and collaboration mechanisms among all stakeholders – ports, terminal operators, shipping and cargo interests, makers of technology, Governments and investors.

## E. OUTLOOK AND POLICY CONSIDERATIONS

In line with projected growth in the world economy, international merchandise trade and seaborne shipments (see chapter 1), prospects for global port-handling activity remain positive overall. The outlook on the supply side is also favourable, as the global port infrastructure market is expected to record the highest gains from 2017 to 2025, primarily owing to increased trade volumes and infrastructural development in emerging developing Asian countries (Coherent Market Insights, 2018).

Energy and container port construction are expected to attract large demand through the forecast period. Western Asia is projected to remain a key investment area, with construction projects such as the Fujairah Oil Terminal, the port and industrial zone of Khalifa (Abu Dhabi), Boubyan Island (Kuwait) and Sohar Industrial Port (Oman), being lined up by the Gulf Cooperation Council. Large-scale projects for fuel handling are also planned in Saldanha Bay (South Africa) and Mombasa (Kenya), as demand for fuels is set to rise with the projected growth of quickly emerging Asian developing countries (Coherent Market Insights, 2018). Port development and refurbishment projects under the Belt and Road Initiative, for example in Pakistan (Gwadar), Djibouti, Myanmar (Kyaukpyu), Greece (Piraeus), and Sri Lanka (Hambantota and Colombo) are contributing to the upgrading and upscaling of port infrastructure in Africa, Asia and Europe. Chinese investment in container ports is expected to grow as port operators in China continue to expand internationally, ultimately surpassing the growth of traditional global operators (Drewry Maritime Research, 2017b).

While overall prospects for global port activity remain positive, preliminary figures are pointing to decelerated growth in port volumes in 2018. This is a reflection of the waning impetus for growth from, in particular, cyclical recovery and supply chain restocking in 2017. Furthermore, downside risks weighing on global shipping, including trade policy risks, geopolitical factors and structural shifts in economies such as China, tend to detract from a favourable outlook. An immediate concern are the trade tensions between China and the United States, the world's two largest economies, and the emergence of inward-looking policies and protectionism (see chapter 1).

Today's overall port-operating landscape is characterized by heightened port competition, especially in containerized trade, where decisions by shipping alliances on capacity deployed and the structure of ports and networks can determine the fate of a container port terminal. Additional investment is required to accommodate larger vessels and larger volumes handled at peak port calls and will likely weigh on port operators' margins (Fairplay, 2017b). However, the cost of new investments could be partially mitigated by exploring tailored pricing to align port and terminal interests with carriers and incentivize shipping lines to work more productively (Port Technology, 2017). Productive and workable cooperative arrangements between port authorities, terminal operators, shipping lines and the trade community will be essential.

When studying the impact of continued market concentration in liner shipping and potential competition concerns, competition authorities and maritime transport regulators should also analyze the impact of market concentration and alliance deployment on the relationship between ports and carriers. Areas of focus include the impact on selection of ports of call, the configuration of liner shipping networks, the distribution of costs and benefits between container shipping and ports, and approaches to container terminal concessions in view of the fact that shipping lines often have stakes in terminal operations.

More than ever, ports and terminals around the world need to re-evaluate their role in global supply and logistics chains and prepare to deal with the changes brought about by the accelerated growth of technological advances with potentially profound impacts (Brümmerstedt et al., 2017). It is important for ports and terminals to seek effective ways to embrace the new technologies to remain competitive and avoid the risk of marginalization in today's highly competitive port industry (Port Equipment Manufacturers Association, 2018).

Enhancing port and terminal performance in all market segments is increasingly recognized as critical for port planning, investment and strategic positioning, as well as for meeting globally established sustainability benchmarks and objectives such as the Sustainable Development Goals. In this context, the port industry and other port stakeholders should work together to identify and enable key levers for improving port productivity, profitability and operational efficiencies. Governments should ensure that policy and regulatory frameworks are supportive and flexible.

Systems that monitor and measure relevant operational, financial and environmental metrics in ports are strategic-planning and decision-making tools that require further support and development. Greater data availability and range enabled by technological advances can be tapped to track, measure and report performance, as well as derive useful insights for port managers, operators, regulators, investors and users. Work carried out under the UNCTAD Port Management Programme on the port performance scorecard could be further developed and its geographical scope expanded.

# REFERENCES

BIMCO (2017). BIMCO's Dry Bulk Terminals Vetting Report for 2017.

Brümmerstedt K, Fiedler R, Flitsch V, Jahn C, Roreger H, Sarpong B, Saxe S and Scharfenberg B (2017). *Digitalization of Seaports: Visions of the Future*. Fraunhofer. Hamburg.

Business Insider Australia (2017). Australia's Port Hedland shipped close to half a billion tonnes of iron ore last financial year. Money and Markets. 7 July.

Clarksons Research (2017). Moving containers globally? Let's stick together. 25 August.

Clarksons Research (2018). *Dry Bulk Trade Outlook*. Volume 24. No. 5. May.

Coherent Market Insights (2018). Port infrastructure market: Global industry insights, trends, outlook, and opportunity analysis, 2016–2024. Press release.

Drewry Maritime Research (2018a). *Container Forecaster*. Quarterly. First quarter.

Drewry Maritime Research (2018b). Ports and terminal insight. Quarterly. First quarter.

Drewry Maritime Research (2017a). *Container Forecaster*. Quarterly. Fourth quarter.

Drewry Maritime Research (2017b). Ports and terminal insight. Quarterly. Fourth quarter.

Fairplay (2017a). Tonnage titans – top 20 ports by annual cargo throughput. 15 October.

Fairplay (2017b). 2017 in review: Port call sizes continue to rise. 15 December.

Fairplay (2018). Improved liner efficiency leaves ports struggling. 3 May.

Hellenic Shipping News (2018). Majority of Navis customers surveyed exploring some level of automation to stay competitive in ocean shipping industry. 15 March.

JOC.com (2018). Global port berth productivity falls as call size continued to grow. 3 May.

Market Realist (2018). What record iron ore shipments from Port Hedland mean for prices. 29 January.

Linked in (2018). Container terminal automation: What does the future really hold? 31 May.

Marine Traffic (2018). Available at www.marinetraffic.com.

Notteboom TE, Parola F, Satta G and Pallis AA (2017). The relationship between port choice and terminal involvement of alliance members in container shipping. *Journal of Transport Geography*. 64:158–173.

Port Equipment Manufacturers Association (2018). Digitalization signals "fourth industrial revolution" for global ports sector. 19 February.

Port Technology (2017). McKinsey report: Vessels to reach 50,000 TEU by 2067. 30 October.

Shanghai International Shipping Institute (2016). Global port development.

Shanghai International Shipping Institute (2017). Global port development.

UNCTAD (2016). *Port Performance: Linking Performance Indicators to Strategic Objectives*. UNCTAD Train for Trade Port Management Series. Volume 4.

UNCTAD (2017a). Port Performance Scorecard Newsletter. Issue 1. https://tft.unctad.org/wp-content/uploads/2017/08/2017-Newsletter-PPS-June-FINAL.pdf.

UNCTAD (2017b). *Information Economy Report 2007: Digitalization, Trade and Development* (United Nations publication, Sales No. E.17.II.D.8, New York and Geneva).

Vonck I (2017). Ports of the future: A vision. Deloitte Port Services. Baltic Ports Conference 2017.

Wei Z (2018). Cosco's presence in Zeebruge fortifies its European Belt and Road. *Shipping and Finance*. February. Issue 260, p. 6.

# 5

# LEGAL ISSUES AND REGULATORY DEVELOPMENTS

Technology has become a crucial element of many systems on board ships and in ports and is continuing to transform and revolutionize the way in which shipping operations are conducted. Many current technological advances, including, for example, autonomous ships, drones and various distributed ledger technologies such as blockchain, hold considerable promise for the increased efficiency of operations and reduced costs, among other possibilities. However, uncertainty remains in the maritime industry with regard to their potential safety and security, and there is concern about the cybersecurity incidents that may occur. To minimize such risks for systems on board ships and in ports, and to facilitate the transition to potential new technologies, Governments and the maritime industry are continuing to improve the safety and risk management culture and making efforts to ensure compliance with the complex and evolving legal framework. In addition, the various distributed ledger technologies currently emerging and proliferating, including blockchain-related initiatives, need to be interoperable, as competition between them in a bid to make a specific technology the chosen standard for the industry may be detrimental for shipping.

As the future of technological advances in shipping is being defined, and the maritime industry is leveraging technology to improve its services, the existing legal, policy and regulatory frameworks are being adapted and new frameworks written, as necessary, at both the national and international levels. The strategic plan for IMO adopted in December 2017 recognizes the need to integrate new and emerging technologies into the regulatory framework for shipping. This plan follows the adoption of a resolution that encourages maritime administrations to ensure that cyberrisks are appropriately addressed in existing safety management systems starting from 1 January 2021, as well as the adoption in July 2017 of the IMO guidelines on maritime cybersecurity risk management.

Important international regulatory developments during the period under review include the adoption by IMO in April 2018 of an initial strategy on the reduction of greenhouse gas emissions from ships, which aims at the reduction of total annual greenhouse gas emissions from ships by at least 50 per cent by 2050, compared with 2008. In addition, IMO adopted a decision with regard to regulatory scoping exercises to establish the extent to which the international regulatory framework should be modified to integrate the new technology involving maritime autonomous surface ships.

This chapter provides a summary of legal and regulatory developments related to these issues and highlights relevant policy considerations for the maritime sector.

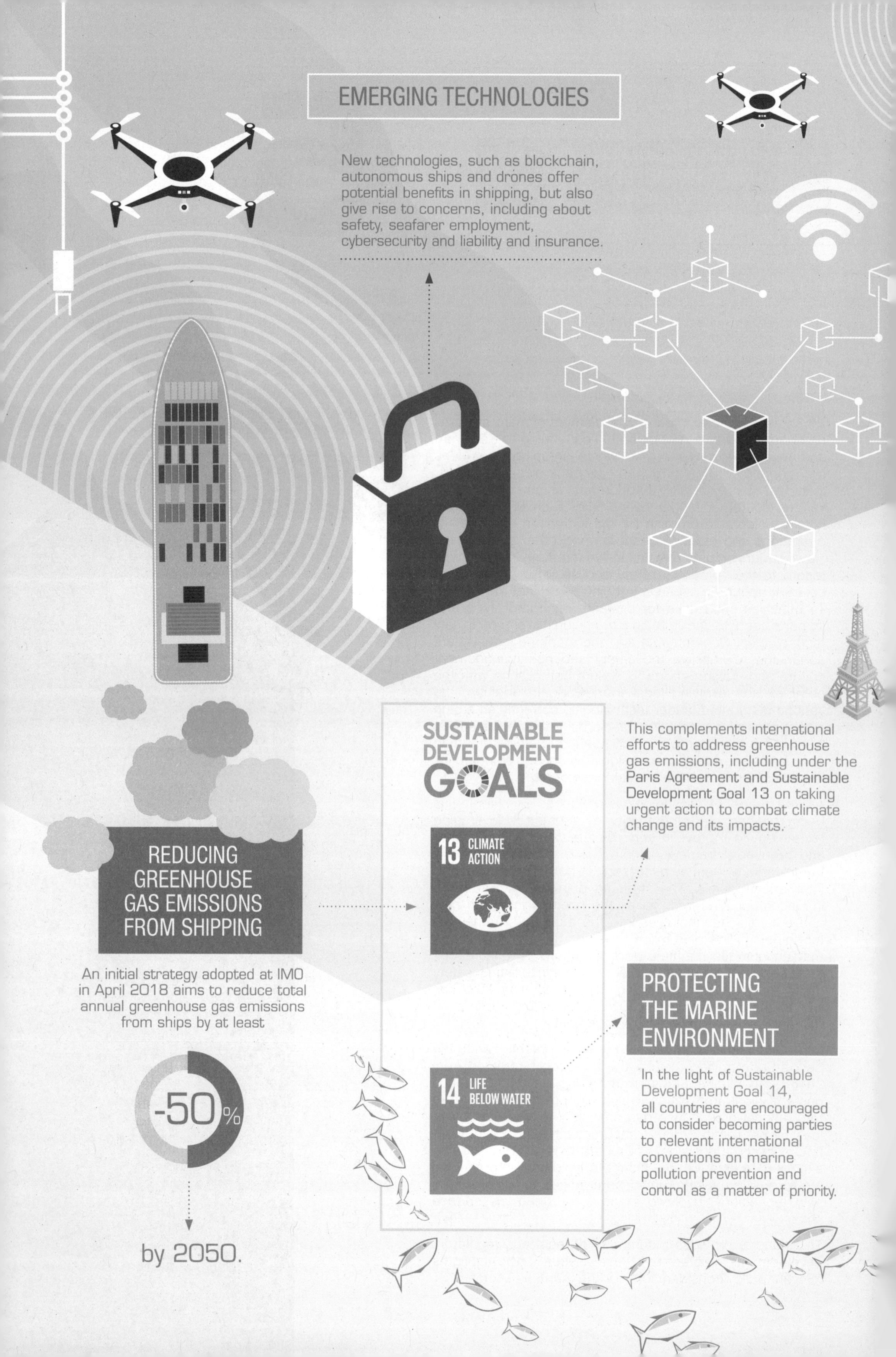
EMERGING TECHNOLOGIES
New technologies, such as blockchain, autonomous ships and drones offer potential benefits in shipping, but also give rise to concerns, including about safety, seafarer employment, cybersecurity and liability and insurance.
REDUCING GREENHOUSE GAS EMISSIONS FROM SHIPPING
An initial strategy adopted at IMO in April 2018 aims to reduce total annual greenhouse gas emissions from ships by at least
-50%
by 2050.
SUSTAINABLE DEVELOPMENT GOALS
13 CLIMATE ACTION
14 LIFE BELOW WATER
This complements international efforts to address greenhouse gas emissions, including under the Paris Agreement and Sustainable Development Goal 13 on taking urgent action to combat climate change and its impacts.
PROTECTING THE MARINE ENVIRONMENT
In the light of Sustainable Development Goal 14, all countries are encouraged to consider becoming parties to relevant international conventions on marine pollution prevention and control as a matter of priority.

## A. TECHNOLOGICAL DEVELOPMENTS AND EMERGING ISSUES IN THE MARITIME INDUSTRY

### 1. Cybersecurity

The *Review of Maritime Transport 2017* highlighted examples of cyberattacks and vulnerabilities in navigation and other systems on board ships and in ports, including interference with automatic identification systems and electronic chart display and information systems, the jamming of global positioning systems and the manipulation of cargo and other ship and port systems, including through the introduction of malware, ransomware and viruses (UNCTAD, 2017a). In particular, 2017 was marked by some major global cyberattacks, including the use of ransomware, which demonstrated that such attacks, although not widely targeted at shipping as yet, may have substantial impacts (*The Guardian*, 2017; ZD Net, 2018). Such incidents and other attacks, including some mass global positioning system-spoofing attacks on ships in the Black Sea, emphasize the importance of cybersecurity and cyberrisk management. Further, there have been reports of links between cyberattacks and physical piracy, whereby pirates have reportedly identified ships with valuable cargo and minimal on-board security by infiltrating the systems of shipping companies.

#### *Cybersecurity guidelines for the maritime industry*

To date, internationally binding cybersecurity regulations for the maritime industry have not been adopted. However, the IMO guidelines on maritime cybersecurity risk management provide high-level recommendations with regard to safeguarding international shipping from current and emerging cybersecurity threats and helping to reduce related vulnerabilities (IMO, 2017a). The guidelines contain five functional elements for effective risk management in the maritime sector, namely to identify, protect, detect, respond and recover (IMO, 2017b). To be effective, these elements need to be incorporated into all aspects of shipping company operations and personnel management, in the same way that the industry has embraced a safety culture, with the adoption of the International Safety Management Code and the implementation of safety management systems. The main purpose of the Code is to provide an international standard for the safe management and operation of ships and for pollution prevention; it establishes safety management objectives and requires the "company", defined as the shipowner or any person, such as the manager or bareboat charterer, who has assumed responsibility for operating a ship, to establish a safety management system and to establish and implement a policy for achieving these objectives (IMO, 2018a). The Maritime Safety Committee of IMO, in its resolution 428(98) on cyberrisk management in safety management systems, encourages administrations to ensure that cyberrisks are appropriately addressed in existing systems as defined in the Code no later than the first annual verification of the company's document of compliance after 1 January 2021. This is the first compulsory deadline established in the maritime industry for cyberrisks and is an important step in protecting the maritime transportation system and the entire maritime industry from increased cybersecurity threats. In addition, the strategic plan for IMO recognizes the need to integrate new and emerging technologies into the regulatory framework for shipping by balancing the benefits derived from such technologies "against safety and security concerns, the impact on the environment and on international trade facilitation, the potential costs to the industry and finally their impact on personnel, both on board and ashore" (IMO, 2017c).

At the same time, the shipping industry is taking a proactive approach to incorporating cyberrisk management into its safety culture, to prevent the occurrence of any serious incidents. Guidance has been and continues to be developed by classification societies and other industry associations. Shortly after the approval of resolution 428(98), industry bodies released the second version of their guidelines on cybersecurity on board ships, which builds on the first version released in 2016 and is more comprehensive. The second version is aligned with the recommendations in the IMO guidelines, provides practical guidance on maritime cyberrisk management and includes information on insurance-related issues. The industry guidelines suggest that cyberrisk management should do the following (BIMCO et al., 2017):

> "Identify the roles and responsibilities of users, key personnel and management both ashore and on board; identify the systems, assets, data and capabilities, which if disrupted, could pose risks to the ship's operations and safety; implement technical measures to protect against a cyberincident and ensure continuity of operations. This may include configuration of networks, access control to networks and systems, communication and boundary defence and the use of protection and detection software; implement activities and plans (procedural protection measures) to provide resilience against cyberincidents. This may include training and awareness, software maintenance, remote and local access, access privileges, use of removable media and equipment disposal; [and] implement activities to prepare for and respond to cyberincidents."

A significant new feature of the second version of the industry guidelines is the fact that they address insurance-related issues with regard to losses from a cybersecurity-related incident. The question of whether such losses should be covered by insurance has to date been unclear. In addressing this issue, the guidelines provide that "companies should be able to demonstrate that they are acting with reasonable care in their approach to managing cyberrisk and protecting the ship from any damage that may arise from a cyber incident" (BIMCO et al., 2017). There is currently no regulation in place on cybersecurity in international shipping, yet maritime companies need to be proactive in addressing cyberrisk, as suggested by IMO and various industry bodies, and can no longer claim ignorance with regard to cyberrisk management.

In addition, the guidelines state that in many markets offering marine property insurance, policies may cover loss or damage to a ship and its equipment caused by a shipping incident such as grounding, collision, fire or flooding, even when the underlying cause of the incident is a cybersecurity-related incident. At present, there are exclusion clauses for cyberattacks in some markets and, if the marine policy contains a relevant exclusion clause, the loss or damage is not covered. In such circumstances, the guidelines recommend that companies verify with insurers and/or brokers in advance with regard to whether the policy covers claims for incidents related to cybersecurity and/or cyberattacks (BIMCO et al., 2017).

More generally, limited data on the frequency of attacks, severity of losses and probability of physical damage remain a challenge to underwriters (All About Shipping, 2018).

Finally, with regard to liability for a cybersecurity-related incident, the guidelines state the following (BIMCO et al., 2017):

> "It is recommended to contact the [protection and indemnity insurance] club for detailed information about cover provided to shipowners and charterers in respect of liability to third parties (and related expenses) arising from the operation of ships. An incident caused, for example by malfunction of a ship's navigation or mechanical systems because of a criminal act or accidental cyberattack, does not in itself give rise to any exclusion of normal [protection and indemnity insurance] cover. It should be noted that many losses which could arise from a cyberincident are not in the nature of third-party liabilities arising from the operation of the ship. For example, financial loss caused by ransomware or costs of rebuilding scrambled data would not be identified in the coverage. Normal cover, in respect of liabilities, is subject to a war risk exclusion and cyberincidents in the context of a war or terror risk, will not normally be covered."

The International Organization for Standardization standard 27001:2013 on information technology – security techniques – information security management systems – requirements, specifies requirements for establishing, implementing, maintaining and continually improving an information security management system within the context of an organization. The standard also includes requirements on the assessment and treatment of information security risks tailored to the needs of the organization. The requirements set out in the standard are generic and intended to be applicable to all organizations, regardless of type, size or nature.

In addition, some countries have also prepared guidelines on cybersecurity. For example, the National Institute of Standards and Technology in the United States published the *Framework for Improving Critical Infrastructure Cybersecurity* in 2018 and the Institution of Engineering and Technology in the United Kingdom published the *Code of Practice: Cybersecurity for Ports and Port Systems* in 2016 and the *Code of Practice: Cybersecurity for Ships* in 2017. Such codes can help companies develop cybersecurity assessments, plans and mitigation measures and manage security breaches, and should be used along with ship security standards and other relevant IMO regulations.

The maritime industry continues to work on improving the understanding of cybersecurity issues and on increasing risk management. Shipping companies are integrating innovative security technologies with existing systems and software, to prevent internal and external cyberattacks with minimal human intervention, including by providing real-time alerts and blocking malicious files to prevent unauthorized access to critical systems and data (Marine Log, 2018).

In addition to verifying that technology, policies and procedures are in place, and that employees at all levels are aware of cyberrisks and how to react in the event of an attack, companies should consider in particular how data is stored and secured, given growing concerns with regard to data usage and security, for example on social media websites, which illustrate the complexity of potential security risks.

Data storage and security is particularly relevant, given the entry into force on 25 May 2018, of European Union Regulation 2016/679 of 27 April 2016 on the protection of natural persons with regard to the processing of personal data and on the free movement of such data, which regulates how companies safeguard the processing and movement of the personal data of citizens of the European Union. Some of the key privacy and data protection provisions of the Regulation include requirements related to the consent of subjects for data processing; anonymization of collected data to protect privacy; provision of data breach notifications; safe handling of the transfer of data across borders; and the appointment by certain companies of a data protection officer to oversee compliance with the Regulation. Notably, it is not only companies in the European Union but any company that processes personal data related to offering goods or services or that monitors the behaviour of European Union residents, regardless of its location, that is subject to the Regulation. In the event of non-compliance, the Regulation provides for the administration of fines by supervisory authorities in member States.

## 2. Internet of things

The Internet of things refers to the network of connected devices with unique identifiers in the form of an Internet protocol address, which have embedded technologies or are equipped with technologies that enable them to sense, gather data and communicate about the environment in which they reside and/or themselves (see www.i-scoop.eu/internet-of-things/).

The shipping sector is increasingly harnessing data generated from satellite information and sensors linking equipment, systems and machinery to support informed decision-making related to route optimization, asset tracking and maintenance. Examples of applications in this domain include software that uses satellite-generated data to determine the most efficient route and estimate in real time the arrival time of vessels; and emerging intelligent containers that use sensors and telematics to track temperature, vibration, humidity and air quality during ocean transport, such as technology used by Maersk and the Mediterranean Shipping Company for reefer monitoring.

The Internet of things is also increasingly used in the industry to improve ship-to-shore connectivity and with regard to intelligent traffic management. A closer interface between ships and ports involves, for example, the use of big data analytics to reduce transit times and time lost when entering ports and other high traffic areas, thereby contributing to alleviating port congestion. For example, the digitalization collaboration initiative between the port of Rotterdam and IBM is helping to prepare this port to host connected ships in future and involves installing sensors across 42 km of land and sea to collect information about traffic management at the port with a view to improving safety and efficiency. A similar initiative between the Maritime and Port Authority of Singapore, academic institutions in Singapore, namely the Institute of High Performance Computing and Singapore Management University, and Fujitsu aims to embed the Internet of things and artificial intelligence technologies to enable long-term traffic forecasts, hotspot calculation and intelligent coordination models.

The Internet of things is also being used to develop systems that support navigation in challenging conditions, such as adverse weather conditions or in congested waterways. For example, in March 2018, Rolls-Royce launched an intelligent awareness system that fuses multiple sensors with intelligent software to create a three-dimensional model of nearby vessels and hazards, to increase safety (Rolls-Royce, 2018). Other applications of the Internet of things currently being tested include the departure of ships without human intervention, the remote controlling of the sailing of ships and the automatic docking of vessels to enable safe berthing (Wärtsilä, 2018).

When shipment events can be recorded in real time, this provides opportunities to optimize operations through blockchain, for example, to track spare capacity, improve connections between different legs of a journey in the global transport network and facilitate capacity-sharing to cope with overcapacity.

## 3. Use of blockchain

Blockchain is a distributed ledger technology that enables peer-to-peer transactions that are securely recorded, as in a ledger, in multiple locations at once and across multiple organizations and individuals, without the need for a central administration or intermediaries. One of the potential problems identified with regard to digital innovation in the maritime industry is insufficient electronic data interchange standardization and the need for a common data format to exchange information (*Combined Transport Magazine*, 2016). Electronic data interchange involves the electronic transfer from one computer to another of commercial or administrative transactions using an agreed standard to structure the transaction or message data (Economic Commission for Europe, 1996). This lack, along with a general lack of clarity with regard to the potential uses of blockchain, are among the factors that may explain the continued reliance in the shipping industry on paper-based documentation for deliveries of cargo containers.

Overall, blockchain holds potential to improve the security of the Internet of things environment. It addresses several aspects of information security, including confidentiality, integrity, availability and non-repudiation. For example, blockchain can protect the security of documents by blocking identity theft, through the use of public key cryptography; preventing data tampering, compared with document signing and other forms of electronic data interchange, through the creation of a public key and a private key; and stopping denial of service attacks, through the removal of the single target that a hacker may attack to compromise an entire system (Venture Beat, 2017). Allowing data to be managed through blockchain could therefore involve adding an extra layer of security and a gradual decrease in the use of centralized storage and processing for data.

In the maritime industry, blockchain has the potential to be used, among others, to track cargo and provide end-to end supply chain visibility; record information about vessels, including on global risks and exposure; integrate smart contracts and marine insurance policies; and digitalize and automate paper filing and documents. Such applications can help save time and reduce costs related to the clearance and movement of cargo. Several initiatives that focus on the container shipping segment have emerged, although blockchain is not yet fully implemented across the sector. Different varieties of maritime single windows are being developed to handle a quotation encompassing an entire ocean transport transaction, including booking, documentation generation and customs clearance. Maritime single windows imply potential efficiency gains and reduced costs for shipping companies due to standardization, which allows fragmented back-end systems to be superseded, and digitalization, which enables the elimination of intermediaries and inefficiencies related to the processing of documentation. For example, Maersk and IBM intend to establish a joint venture, which remains subject to the receipt of regulatory approvals. The aim of the venture is to develop an open trade-digitalization platform, designed for use by the entire industry, to help companies move and track goods digitally across international borders. The platform will use blockchain and other cloud-based, open-source technologies, including artificial intelligence, the Internet of things and analytics, delivered through IBM, and initially commercialize the following two core capabilities aimed at digitalizing the global supply chain (Maersk, 2018):

> "A shipping information pipeline will provide end-to-end supply chain visibility to enable all actors involved in managing a supply chain to securely and seamlessly exchange information about shipment events in real time; paperless trade will digitize and automate paperwork filings by enabling end users to securely submit, validate and approve documents across organizational boundaries, ultimately helping to reduce the time and cost for clearance and cargo movement. Blockchain-based smart contracts ensure all required approvals are in place, helping speed up approvals and reducing mistakes."

Another example of the use of blockchain in shipping is the completion by Hyundai Merchant Marine and other members of a consortium, in September 2017, of a pilot voyage applying blockchain that used secure paperless processes for shipment booking and cargo delivery. Hyundai Merchant Marine also reviewed the feasibility of introducing the technology into shipping and logistics and tested and reviewed the combination of blockchain with the Internet of things through the real-time monitoring and management of the reefer containers on the vessel (Lloyd's List, 2017).

In addition, in August 2017, Japan formed a consortium of 14 members to develop a platform for sharing trade data using blockchain, and Singapore-based Pacific International Lines signed a memorandum of understanding with PSA International and IBM in Singapore to develop and test supply chain business network solutions based on blockchain (Lloyd's List, 2017). Other initiatives include the cargo-booking portals of INTTRA and GT Nexus; the e-commerce business platform of CMA CGM; and the single window at the port of Cotonou, facilitated by the World Bank, to ease the management of vessel traffic, cargo and intermodal operations.

Potential future applications of blockchain in shipping could include smart contracts, which are contracts in the form of a computer programme run within blockchains that automate the implementation of the terms and conditions of any agreement between parties. Several smart contract prototypes have been launched that involve digitalizing electronic bills of lading and other trade documents, such as CargoDocs under essDOCS and Cargo X. However, the development of financing, payment and insurance aspects related to shipping remain in experimental and pilot stages. Once the use of such contracts reaches maturity, possible scenarios include the negotiation of freight prices directly between asset owners and their counterparts; the automatic processing of payments upon specified conditions being satisfied; and the issuance of insurance policies and settling of marine insurance claims through blockchain.

Blockchain has been deployed for the first time in the marine insurance sector. In May 2018, some industry actors collaborated with Ernst and Young and the software security firm Guardtime to launch the world's first blockchain-based platform for marine hull insurance. The platform, which is ready for commercial use, is expected to help manage risk for more than 1,000 commercial vessels in its first year and is planned to be implemented for other types of insurance for the marine cargo, global logistics, aviation and energy sectors (Splash 247, 2018). The platform "connects clients, brokers, insurers and third parties to distributed common ledgers that capture data about identities, risk and exposures and integrates this information with insurance contracts" and has the ability to "create and maintain asset data from multiple parties; to link data to policy contracts; to receive and act upon information that results in a pricing or a business process change; to connect client assets, transactions and payments; and to capture and validate up-to-date first notification or loss data" (Guardtime, 2017).

In addition, in 2017, two logistics companies, along with a containership operating company, completed a pilot project on blockchain-based paperless bills of lading that involved the use of an application for the issuance, transfer and reception of original electronic documents, and the containers, shipped from China to Canada, were successfully delivered to the consignees (Marine Log, 2017). The potential use of blockchain in this context is worth noting, as commercially viable electronic alternatives to traditional paper-based bills of lading have only recently emerged. Earlier attempts in this regard include the Bill of Lading Electronic Registry Organization (UNCTAD, 2003; www.bolero.net) and, more recently and with some success, essDOCS (www.essdocs.com). The main challenge in efforts to develop electronic alternatives to traditional paper-based transport documents has been the effective replication of a document's functions in a secure electronic environment while ensuring that the use of electronic records or data messages has the same legal recognition as that of paper documents. With regard to bills of lading, as the exclusive right to the delivery of goods has traditionally been linked to the physical possession of original documents, this includes in

particular the replication in an electronic environment of the unique document of title function (UNCTAD, 2003).

Blockchain is also being used to improve tuna traceability to help end illegal and unsustainable fishing practices in the tuna industry in Asia and the Pacific. In January 2018, the World Wide Fund for Nature in Australia, Fiji and New Zealand, in partnership with a technology innovator, a technology implementer and a tuna fishing and processing company, launched a pilot project in the tuna industry in the Pacific that will use blockchain to track the journey of tuna "from bait to plate", strengthening transparency and traceability. The aim is to help end illegal, unreported and unregulated fishing and human rights abuses of seafarers and workers in the tuna industry and to address safety issues and broader impacts on the environment (The Conversation, 2018a).

Finally, blockchain is also proliferating in terminal and port development. For example, in April 2015, construction was completed of a fully automated and environmentally sustainable container terminal at the port of Rotterdam, and in September 2017, a field laboratory, Block Lab, was launched, which is aimed at developing applications and solutions based on blockchain.

Given that many blockchain initiatives and partnerships are proliferating, there is a need for the different applications emerging in the shipping industry to be interoperable. As noted by observers, "it would be detrimental for the shipping industry if the different factions and initiatives compete head on trying to make their specific blockchain technology choice the de facto standard for the industry" (JOC.com, 2018). Blockchain promises secure transactions yet, according to some specialists, it may not be as secure as generally anticipated. The use of blockchain may help solve some security issues but may also lead to new, potentially more complex security challenges, as some methods can possibly still be used to hack into a maritime transaction blockchain, including compromising the private keys of users; cracking cryptography, given continuous advances in computing; obtaining control of a majority of the mining nodes used to implement blockchain; and abusing vulnerabilities in smart contracts or coded programmes supported and run within blockchains (Marine Electronics and Communications, 2018a).

There are also concerns that many developing countries, in particular the least developed countries, may be inadequately prepared to capture the opportunities and benefits emerging from digitalization. There is a risk that digitalization may lead to increased polarization and widening income inequalities, as productivity gains might accrue mainly to a few, already wealthy and skilled individuals, given that "winner-takes-all dynamics are typical in platform-based economies, where network effects benefit first movers and standard setters" and that "the overall effects of digitalization remain uncertain; they will be context-specific, differing greatly among countries and sectors [and this] makes it increasingly important for countries to ensure they have an adequate supply of skilled workers with strong non-cognitive, adaptive and creative skills necessary for 'working with the machines'" (UNCTAD, 2017b). Additional concerns have been raised about digitalization, as it could potentially lead to a fragmentation of the global provision and international trade of services. This could open up new avenues for the development strategies of developing countries, yet it is unclear whether digital-based services could provide similar employment, income and productivity gains as manufacturing has traditionally done; "disruptive technologies always bring a mix of benefits and risks [but] whatever the impacts, the final outcomes for employment and inclusiveness are shaped by policies" (UNCTAD, 2017c).

## 4. Autonomous ships, drones and other innovations in shipping

### *Autonomous ships: Potential benefits and challenges*

Among the advances in cybersystems and digitalization in the maritime industry, maritime autonomous surface ships, also known as unmanned surface vessels, are attracting increased attention. As with autonomous technologies in other industries, autonomous ships have the potential to provide enhanced safety and cost savings by removing the human element from certain operations. The term "autonomous ship" is not the same as "unmanned ship", as the former may operate at various levels of autonomy, including partially autonomous (with human input) and fully autonomous (not requiring human intervention). However, such terms have not yet been completely defined either nationally or internationally, and many different formulations exist of the levels of autonomy (Danish Maritime Authority, 2017). In any event, human intervention will still be needed in most ship operations in the near future, and the transportation of cargo and passengers in fully autonomous ships remains a long-term ambition. Autonomous ships could potentially be used in a wide range of operations, including salvage, oil spill response, passenger ferrying, offshore supply, towing and the carriage of cargo. However, at present, they are mostly used for marine scientific research and various maritime operations in the defence sector (Comité Maritime International, 2017). The first remotely controlled or fully autonomous commercial cargo vessel may be in operation by 2020; for example, the first fully electric and autonomous container ship, with zero emissions, may be in operation on a short coastal route in either a remotely controlled or autonomous mode by 2020 (Marine Electronics and Communications, 2018b). The technology may first be deployed on vessels that undertake coastal and short sea routes, and remotely controlled and autonomous ships sailing open oceans could be in operation by 2030 or earlier. An autonomous, fully battery-powered short sea vessel with zero emissions is also currently in development (DNV GL, 2018).

Other recent developments with regard to autonomous ships include the following: a prototype of the world's first fully autonomous and cost-efficient vessel for offshore operations (Kongsberg, 2017); the first electrically powered inland container vessel in Europe, with five small ships in the series expected to be completed in 2018 and six larger ships in preparation with features that prepare them for autonomous operations (*The Maritime Executive*, 2018); an agreement between two companies, possibly a first in the marine sector, to develop an artificial intelligence-based classification system for detecting, identifying and tracking the objects a vessel can encounter at sea, aimed at making existing vessels safer and progressing towards making autonomous ships a reality (Rolls-Royce, 2017); the One Sea autonomous maritime ecosystem project, aimed at enabling fully remote-controlled vessels in the Baltic Sea by 2020 and achieving autonomous commercial operations by 2025 (IMO, 2018b); and the testing of remotely controlled vessels in the Pacific Ocean, due to begin in 2019, aimed at achieving autonomous vessels by 2025 (Bloomberg, 2017).

An area that might benefit from the use of autonomous ships is the safety and security of ship operations. Advances have been made in electronic navigational systems and tools, yet the human factor continues to have an important role in most marine incidents and casualties. Some studies estimate that 75–96 per cent of marine accidents can be attributed to human error and human error reportedly accounted for approximately 75 per cent of the value of almost 15,000 marine liability insurance claims in 2011–2016, equivalent to over $1.6 billion (Allianz Global Corporate and Specialty, 2017).

Crew costs can constitute up to 42 per cent of a ship's operating costs (Stopford, 2009). This cost decreases for vessels with fewer or no crew, as does the risk of piracy and hostage-taking and the respective insurance coverage rates and costs. Vessel construction costs may also be reduced, with less space required for seafarer accommodation and other amenities, which could instead be used for cargo storage. Vessel operations could also become more environmentally friendly, as new autonomous ships are designed to operate with alternate fuel sources, zero-emissions technologies and no ballast. In addition, given fewer or no crew on board, there would be less garbage and sewage to manage and treat.

There are a number of potential benefits, yet challenges in implementation, which include concerns about the following: cybersecurity, although this is not unique to autonomous ships; safety, related to the lack of crew on board; undue impacts on seafarer jobs and shipping rates; and whether insurance cover would be offered by underwriters, insurers and protection and indemnity insurance clubs for commercial autonomous ships (Fairplay, 2017). The potential loss of seafarer jobs is a particular concern in developing countries, as a significant majority of seafarers are from these countries.

### *Autonomous ships: Regulatory issues*

The operation of autonomous ships is closely related to the roles of master and crew on board, a feature that affects the full spectrum of applicable maritime laws and regulations. Regulatory frameworks governing the maritime industry have had to adapt over the years to accommodate new technologies, yet they do not take into consideration the operation of ships without a crew. Therefore, the traditional on-board roles of master and crew, as well as artificial intelligence and shore-based staff supervising remotely controlled or autonomous ships will need to be assessed and redefined. At the international level, aspects of the regulatory framework that need to be considered in the context of autonomous ships include the following:

- Jurisdictional rules specifying the rights and obligations of States with regard to ships in various marine areas and, more specifically, the principles and rules related to flag, port and coastal State jurisdictions, which are mostly covered by the United Nations Convention on the Law of the Sea, 1982. This is a widely ratified framework convention, with 168 States Parties as at 31 July 2018, which defines the rights and responsibilities of nations with regard to their use of the world's oceans, the protection of the marine environment and the management of marine natural resources.
- Technical rules related to, among others, safety, security and the environment, seafarer issues, training and watchkeeping standards, which impose obligations on flag States to enact national legislation reflecting the internationally agreed standards developed by and adopted at IMO.
- Private law rules covering liability for, among others, personal injury, pollution, cargo-related losses and collisions, which are in some instances subject to relevant international legal instruments but may also be subject to national laws.

Recent international regulatory developments of note include a scoping exercise for the review of relevant instruments, to ensure the safe design, construction and operation of autonomous ships, initiated at IMO in 2017 following a decision by the Maritime Safety Committee. A similar review was proposed by the Legal Committee in April 2018, aimed at ensuring that the legal framework set out in legal instruments under its purview provides for the same level of protection for autonomous ships as that provided for operations with non-autonomous ships (IMO, 2018b). Other committees, including the Facilitation Committee and the Marine Environment Protection Committee, may need to undertake similar reviews, as some of the IMO instruments that may need to be considered as part of a comprehensive regulatory review fall under their purview. The Technical Cooperation Committee may also have inputs, in particular when implementation issues are considered.

A cross-divisional task force has been established to facilitate the coordination of work between different committees (IMO, 2018c; IMO, 2018d). In May 2018, the Maritime Safety Committee requested the IMO secretariat to review the work undertaken to date by several organizations that had considered regulatory arrangements and submitted the results of their work to the Committee, and to submit a consolidated report for its consideration at its 100th session in December 2018 (IMO, 2018d; for further information, see the following documents: MSC 99/5, MSC 99/5/1-12, MSC 99/INF.3, MSC 99/INF.5, MSC 99/INF.8, MSC 99/INF.13, MSC 99/INF.14 and MSC 99/INF.16).

Some of the most pertinent IMO instruments with requirements that may need to be evaluated in the context of the navigation of autonomous ships are addressed in this section.

### International Convention for the Safety of Life at Sea, 1974

This Convention is the most important of all of the international conventions concerning the safety of commercial ships, and is widely ratified, with 164 States Parties as at 31 July 2018. It applies to over 99 per cent of the world's tonnage and specifies the minimum standards for the construction, equipment and operation of ships, compatible with their safety. This Convention is one of the key IMO conventions, along with the International Convention for the Prevention of Pollution from Ships, 1973/1978, and the International Convention on Standards of Training, Certification and Watchkeeping for Seafarers, 1978, as amended. In addition, the Maritime Labour Convention, 2006, with 88 ratifications as at 31 July 2018, and representing 91 per cent of the world's tonnage, is the main international instrument setting out seafarers' rights to decent conditions of work. These Conventions constitute the four pillars of the international regulatory regime for quality shipping.

A review of 12 chapters of the International Convention for the Safety of Life at Sea, as follows, may be needed to determine how autonomous ships may be covered by the provisions: chapter I, general provisions, including definitions; chapter II-1, construction, including structure, subdivision and stability, machinery and electrical installations; chapter II-2, fire protection, fire detection and fire extinction; chapter III, life-saving appliances and arrangements; chapter IV, radiocommunications; chapter V, safety of navigation; chapter VI, carriage of cargoes; chapter VII, carriage of dangerous goods; chapter VIII, nuclear ships; chapter IX, management for the safe operation of ships; chapter X, safety measures for high-speed craft; chapter XI-1, special measures to enhance maritime safety; and chapter XII, additional safety measures for bulk carriers.

For example, a review of relevant provisions in chapter V on the safety of navigation may be particularly relevant, as some of the provisions require that, from the point of view of safety, all ships must be sufficiently and efficiently staffed. Other provisions relate to the establishment of control of a ship in hazardous navigational situations and the obligation for the master of a ship to provide assistance to persons in distress at sea. A ship operating autonomously without any human oversight would not be able to comply with such provisions and, should an incident occur, issues related to safety and liability might arise. Such functions may have to be taken over by shore-based staff supervising remote-controlled or autonomous ships, and many of the liabilities may have to be assumed by shipowners, shipbuilders and manufacturers of ship components, as has been addressed in similar situations involving autonomous vehicles (The Conversation, 2018b). A way of apportioning responsibility between these parties and third parties needs to be identified, as existing liability rules applicable in the context of traditional staffed maritime activity cannot be simply transplanted to autonomous counterparts.

The provisions in chapter XI on special measures to enhance maritime safety are also particularly relevant, as they require compliance with the International Ship and Port Facility Security Code, and deal with, among others, the specific obligations of ship companies with regard to security, including security procedures, the employment of security-focused personnel and certification and verification requirements. The unique security challenges posed in the context of autonomous operability are relevant in this regard, in particular with regard to cyberinfiltration. Regulation 6 in this chapter requires ships to have a security alert system that transmits ship-to-shore security alerts to designated authorities that indicate the location of a ship and that its security is under threat, which must be able to be engaged from the bridge and at least one other location. A similar alert mechanism might therefore need to be established in an autonomous ship. Regulation 8 requires that the discretion of a master not be constrained by the company or any other person in respect of ship safety. In an autonomous ship, this role might need to be transferred to a shore-based remote controller.

### International Regulations for Preventing Collisions at Sea, 1972

The Regulations set out navigational rules to be followed by vessels, aimed at avoiding collisions. A review of the five parts, as follows, may be needed to determine how autonomous ships may be covered: part A, general, including provisions related to applicability; part B, steering and sailing; part C, lights and shapes; part D, sound and light signals; and part E, exemption.

### International Convention on Standards of Training, Certification and Watchkeeping for Seafarers

The Convention as amended prescribes qualification standards for masters, officers and watchkeeping personnel on board seagoing ships, along with watchkeeping procedures. Article 3, for example, specifies that the Convention applies to seafarers serving on board seagoing ships entitled to fly the flag of a State Party. The provisions would therefore need to be amended before they could apply to autonomous ships.

#### International Convention for the Prevention of Pollution from Ships

This Convention is the main international convention covering the prevention of pollution of the marine environment by ships from operational or accidental causes and is widely ratified, with 157 States Parties as at 31 July 2018, and applies to over 99 per cent of the world's tonnage. It includes six technical annexes, as follows: annex I, regulations for the prevention of pollution by oil; annex II, regulations for the control of pollution by noxious liquid substances in bulk; annex III, prevention of pollution by harmful substances carried by sea in packaged form; annex IV, prevention of pollution by sewage from ships; annex V, prevention of pollution by garbage from ships; and annex VI, prevention of air pollution from ships.

Autonomous ships, when in operation, would have to comply with relevant provisions in the Convention to the same extent as traditional staffed vessels including, among others, provisions with regard to construction and equipment-related requirements for various types of ships such as oil tankers; operational and procedural requirements such as discharge limits and ship-to-ship transfers; and reporting requirements in the event of spills. These provisions will therefore need to be reviewed.

#### Paris Memorandum of Understanding on Port State Control, 1982

This Memorandum was concluded by 14 European shipping nations and aims to ensure an effective system for controlling the technical condition and safety of ships, in addition to inspections by the flag State. The Memorandum was also motivated by the fact that a number of flags of convenience had historically proven to not be able to effectively control ships flying their flags. The Memorandum establishes a system for port State control of ships from all countries calling at a port in States Parties. At present, the Memorandum covers all member States of the European Union, as well as Canada, Iceland, Norway and the Russian Federation, and the United States is affiliated as a cooperating country. Port State control under the Memorandum includes the inspection of seafarer certificates of competency and qualifications according to the International Convention on Standards of Training, Certification and Watchkeeping for Seafarers, as well as compliance with the International Convention for the Safety of Life at Sea, the International Convention for the Prevention of Pollution from Ships and the Maritime Labour Convention. Inspired by the Memorandum, similar regional port State control agreements have been concluded in Asia and the Pacific and in Latin America. In the European Union, Directive 2009/16 of 23 April 2009 on port State control, based on the Memorandum, sets out a number of additional obligations for information exchanges and reporting between member States of the European Union with regard to port State control, as well as on the professional qualifications of ship surveyors. Such instruments will also need to be reviewed with regard to autonomous ships.

Examples of international legal instruments and legal issues that the Legal Committee of IMO may need to examine with regard to autonomous ships are outlined below.

#### Nairobi International Convention on the Removal of Wrecks, 2007

This Convention, with 41 States Parties as at 31 July 2018, representing 72.41 per cent of the world's tonnage, provides the legal basis for States to remove or have removed shipwrecks that may have the potential to adversely affect the safety of lives, goods and property at sea, as well as the marine environment. With regard to autonomous ships, the terms "master" and "operator" and the requirement for the master and operator of a ship to report a wreck may need to be reviewed. In addition, the requirement that the master and operator report without delay on the nature of the damage may need to be reviewed. The requirement under various liability conventions that certificates attesting that insurance or other financial security is in place must be carried on board may not be relevant if there is no crew on board (IMO, 2018b).

#### Other relevant instruments

Other relevant instruments that may be covered under the scoping exercise include the following: Convention on Facilitation of International Maritime Traffic, 1965; International Convention on Load Lines, 1966; International Convention on Tonnage Measurement of Ships, 1969; International Convention on Maritime Search and Rescue, 1979; Convention for the Suppression of Unlawful Acts Against the Safety of Maritime Navigation, 1988; and International Convention on Salvage, 1989.

#### *Autonomous ships: Jurisdictional issues*

According to the United Nations Convention on the Law of the Sea, which in large part codifies established customary international law, the nationality of a ship is determined by its flag, that is, by its country of registration, and the law of the flag State applies to the ship or any conduct that takes place on it (articles 91 and 94). Each State has the right to determine the conditions for granting its nationality to ships, for registering ships in its territory and for the right to fly its flag (article 91 (1)), as well as the obligation to maintain a register of ships flying its flag (article 94 (2) (a)). Flag States have an important role in the implementation and enforcement of international conventions, including those dealing with the technical and safety aspects of shipping, seafarer working conditions and crew training, and in monitoring compliance with relevant mandatory standards (article 94). In parallel with flag State jurisdiction, which applies to a ship irrespective of its location, port and coastal State jurisdiction also applies, depending on the maritime zone in which the ship is located, that is, a port, internal waters, a territorial sea, an exclusive economic zone or the high seas (Comité Maritime International, 2017).

### *Autonomous ships: Definitions*

Certain concepts such as master and crew and related qualifications that may already exist in various international conventions that presume there is a crew on board, such as article 94 (4) (b) of the United Nations Convention on the Law of the Sea, may need to be clarified with regard to their applicability to autonomous ships. The definition of the terms "vessel" and "ship" may also need to be reviewed, as they may exist in various international conventions based on their area of focus, such as the Nairobi International Convention on the Removal of Wrecks, the International Convention on Salvage and the International Convention on Civil Liability for Oil Pollution Damage, 1969, and its 1992 Protocol.

### *Autonomous ships: Liability rules*

Liability rules applicable in the context of traditional staffed maritime activity cannot be applied to the various levels of autonomy in the context of autonomous ships. New regulations and practices may need to be developed that will likely "involve further standards of due diligence on the part of the shipowner, additional certification requirements for component/software developers and new training and qualification standards for pre-programming and shore-based navigation" (Comité Maritime International, 2017).

### *Drones*

Drones, that is, unmanned aircraft, may offer benefits to the maritime industry with regard to, for instance, cost reduction, the saving of time and the enhancement of safety for operations traditionally conducted by staff. A number of companies are developing autonomous drones to enable the following: inspect and survey ships and offshore installations (DNV GL, 2017; UASweekly.com, 2018); map oil spills and assist in rescue operations (see, for example, www.planckaero.com/maritimedrone); monitor emissions from ships (SUAS News, 2017); and carry and deliver goods and supplies (Baird Maritime, 2018; Fast Company, 2017; *The Maritime Executive*, 2017). However, the relevant jurisdictional issues and implications for the legal framework governing combined aviation and maritime operations need to be further explored and better understood.

# B. REGULATORY DEVELOPMENTS RELATED TO THE REDUCTION OF GREENHOUSE GAS EMISSIONS FROM INTERNATIONAL SHIPPING AND OTHER ENVIRONMENTAL ISSUES

## 1. Reduction of greenhouse gas emissions

Carbon dioxide emissions from international shipping have increasingly been in the spotlight, in particular as they are not covered under the 1997 Kyoto Protocol to the United Nations Framework Convention on Climate Change. Relevant regulations have been considered under the auspices of IMO, including the adoption in 2011 of a set of technical and operational measures to reduce emissions from international shipping and related guidelines (UNCTAD, 2011a; UNCTAD, 2012a). More recently, following the adoption in 2015 of the Paris Agreement under the Convention, further progress has been made, including the adoption in 2016 of a road map for developing a comprehensive IMO strategy on the reduction of greenhouse gas emissions from ships (IMO, 2016, annex 11), and the adoption of an initial strategy in 2018.

### *Initial strategy on greenhouse gas emissions*

According to IMO estimates, in 2012, greenhouse gas emissions from international shipping accounted for 2.2 per cent of anthropogenic carbon dioxide emissions and relevant emissions could increase by between 50 and 250 per cent by 2050 (IMO, 2014). This is of particular concern, given the internationally agreed goal in the Paris Agreement of limiting the global average temperature increase to below 2°C above pre-industrial levels, which will require worldwide emissions to be at least halved from the 1990 level by 2050. The implementation of technical and operational measures for ships could increase efficiency and reduce emissions by up to 75 per cent and further reductions could be achieved by implementing innovative technologies (IMO, 2009).

In April 2018, the seventy-second session of the Marine Environment Protection Committee, at a meeting attended by more than 100 member States of IMO, adopted an initial strategy on the reduction of greenhouse gas emissions from ships (IMO, 2018e). The strategy envisions reducing greenhouse gas emissions from international shipping and phasing them out as soon as possible before 2100. This complements international efforts to address greenhouse gas emissions, including under the Paris Agreement and the 2030 Agenda for Sustainable Development, in particular Sustainable Development Goal 13 on taking urgent action to combat climate change and its impacts. In addition, the strategy sets out relevant guiding principles, including the principles of non-discrimination and of no more favourable treatment, as enshrined in the International Convention for the Prevention of Pollution from Ships and other IMO conventions, as well as the principle of common but differentiated responsibilities and respective capabilities, in the light of different national circumstances, as enshrined in article 4 of the United Nations Framework Convention on Climate Change, including the Kyoto Protocol and the Paris Agreement. The strategy identifies candidate short-term, midterm and long-term further measures, with possible timelines and their impacts on States, stating that specific attention should be paid to the needs of developing countries, in particular the least developed countries and small island developing States. It also identifies supportive measures, including capacity-building, technical cooperation and research and development.

According to the 2016 road map, a revised strategy is to be adopted in 2023. Under short-term measures to be further developed and agreed upon by member States in 2018–2023, the initial strategy includes technical and operational energy efficiency measures for both new and existing ships, including for speed optimization and reduction, and the use of alternative low-carbon and zero-carbon fuels for marine propulsion and other new technologies. Under midterm measures to be agreed upon in 2023–2030, the strategy includes innovative emissions-reduction mechanisms, possibly including market-based measures, to incentivize the reduction of greenhouse gas emissions. Under long-term measures to be undertaken beyond 2030, the strategy aims for measures that will lead to zero-carbon or fossil-free fuels, to enable the potential decarbonization of the shipping sector after 2050. The strategy notes that "technological innovation and the global introduction of alternative fuels and/or energy sources for international shipping will be integral" to achieving the overall ambition, and includes the following levels of ambition (IMO, 2018f, annex 1):

> "1. Carbon intensity of the ship to decline through implementation of further phases of the energy efficiency design index for new ships: to review with the aim to strengthen the energy efficiency design requirements for ships with the percentage improvement for each phase to be determined for each ship type, as appropriate; 2. carbon intensity of international shipping to decline: to reduce [carbon dioxide] emissions per transport work, as an average across international shipping, by at least 40 per cent by 2030, pursuing efforts towards 70 per cent by 2050, compared to 2008; and 3. [greenhouse gas] emissions from international shipping to peak and decline: to peak [greenhouse gas] emissions from international shipping as soon as possible and to reduce the total annual [greenhouse gas] emissions by at least 50 per cent by 2050 compared to 2008 whilst pursuing efforts towards phasing them out as called for in the vision as a point on a pathway of [carbon dioxide] emissions reduction consistent with the Paris Agreement temperature goals."

### *Energy efficiency*

Energy efficiency measures have been legally binding in the maritime industry since 2013, following the entry into force of relevant amendments to annex VI of the International Convention for the Prevention of Pollution from Ships, and include the energy efficiency design index, which sets standards for new ships and associated operational energy efficiency measures for existing ships. In April 2018, the Marine Environment Protection Committee was advised that nearly 2,700 new ships had been certified as complying with energy efficiency standards, and adopted amendments to annex VI, regulation 21 on energy efficiency design index requirements for roll-on roll-off cargo and passenger ships (IMO, 2018e). A correspondence group is expected to present an interim report in October 2018 and a final report in 2019 with recommendations on the time periods and reduction rates for requirements for phase 3 of the energy efficiency design index and the possible introduction of requirements for phase 4. In addition, amendments to the Convention have entered into force that make a data collection system for the fuel oil consumption of ships of 5,000 gross tons and above mandatory, with data collection from 1 January 2019. The data must be reported to the flag State after the end of each calendar year and subsequently transferred to the IMO database.

In addition to technical and operational measures, discussions on market-based measures to reduce emissions from international shipping have been ongoing at IMO, yet an agreement has not yet been reached (UNCTAD, 2011a; UNCTAD, 2012a; for a summary of potential market-based measures currently under discussion, see chapter 3). In 2013, formal discussions on market-based measures at the Marine Environment Protection Committee were suspended (IMO, 2013). The topic was considered at meetings of the Intersessional Working Group on Reduction of Greenhouse Gas Emissions from Ships in June and October 2017 with regard to its possible inclusion in a strategy on the reduction of emissions (IMO, 2017d; IMO, 2017e). The reports of the meetings reflect the different views expressed, in particular that measures "will include technical and operational measures, but market-based measures may be needed in the medium term whilst alternative fuels are developed" and that "market-based measures should be addressed as candidate midterm measures in order to help incentivize uptake of alternative fuels; potentially market-based measures can be designed not to only remove funds from the sector but also to bring funds into the sector to support greater emissions reductions" (IMO, 2017d; IMO, 2017e). The initial strategy on the reduction of emissions from ships includes among candidate midterm measures new and innovative emission-reduction mechanisms, possibly including market-based measures, to incentivize the reduction of greenhouse gas emissions (IMO, 2018f).

## 2. Ship-source pollution and protection of the environment

Other recent regulatory developments under the auspices of IMO regarding ship-source pollution control and environmental protection, aimed at ensuring clean and environmentally sustainable shipping, cover air pollution, ballast water management, hazardous and noxious substances and marine litter.

### *Air pollution*

Sulphur oxides and nitrogen oxides, through chemical reactions in the air, are converted into fine particles that, in addition to particles directly emitted by ships such as black carbon and other carcinogenic particles, increase the health-related impacts of shipping pollution and are linked to premature deaths. The Review of Maritime Transport 2017 noted that an important decision had been adopted at IMO, whereby the global limit of 0.5 per cent on sulphur in fuel oil, as set out in annex VI, regulation 14.1.3 of the International Convention for the Prevention of Pollution from Ships, would come into effect on 1 January 2020 (UNCTAD, 2017a). Within emission control areas in which more stringent controls on sulphur oxide emissions apply, the sulphur content of fuel oil must be no more than 0.1 per cent (1,000 parts per million) from 1 January 2015. The first two sulphur oxide emission control areas were established in Europe, in the Baltic Sea and the North Sea, and took effect in 2006 and 2007, respectively; the third was established in North America and took effect in 2012; and the fourth was established as the United States Caribbean Sea, covering waters adjacent to the coasts of Puerto Rico and the United States Virgin Islands, and took effect in 2014. The consistent implementation of a global sulphur content limit for all ships is expected to bring positive results for human health and the environment, in particular as shipping emissions are associated with a large number of fatalities and illnesses at the global level (Independent, 2018).

In April 2018, the Marine Environment Protection Committee approved draft amendments to annex VI of the International Convention for the Prevention of Pollution from Ships, concerning the prohibition on the carriage of non-compliant fuel oil, with sulphur content exceeding 0.5 per cent, for combustion purposes for propulsion or operation on board a ship (IMO, 2018e). Ships fitted with an approved equivalent arrangement to meet the sulphur limit, such as an exhaust gas cleaning system or scrubber, permitted under annex VI, regulation 4.1, would be exempt. Under regulation 3.2, ships undertaking research trials of emissions reduction and control technology could also be exempt. Guidelines to support the implementation of the sulphur limit to come into effect on 1 January 2020 are in preparation at IMO. Finally, the Committee approved guidance on best practices for fuel oil purchasers and users for assuring the quality of fuel oil used on board ships.

### *Ballast water management*

A significant achievement in 2017 was the entry into force on 8 September of the International Convention for the Control and Management of Ships' Ballast Water and Sediments, 2004. As at 31 July 2018, the Convention had 75 States Parties, representing 75.34 per cent of the world's tonnage. The Convention aims to prevent the risk of the introduction and proliferation of non-native species following the discharge of untreated ballast water from ships. This is considered one of the four greatest threats to the oceans and one of the major threats to biodiversity that, if not addressed, could have severe public health-related, environmental and economic impacts (UNCTAD, 2011b; UNCTAD, 2015; see http://globallast.imo.org). From 8 September 2017, ships are required to manage their ballast water to meet standards referred to as D-1 and D-2; the former requires ships to exchange and release at least 95 per cent of ballast water by volume far away from a coast and the latter raises the restriction to a specified maximum amount of viable organisms allowed to be discharged, limiting the discharge of specified microbes harmful to human health. In April 2018, the Marine Environment Protection Committee adopted amendments to the Convention that clarify when ships must comply with the D-2 standard. New ships, constructed on or after 8 September 2017, shall meet the D-2 standard from the date they enter into service. Existing ships constructed before 8 September 2017 shall comply with the D-2 standard after their first or second five-year renewal survey associated with the International Oil Pollution Prevention Certificate under annex I of the International Convention for the Prevention of Pollution from Ships conducted after 8 September 2017, and in any event not later than 8 September 2024 (IMO, 2017f). Given the entry into force of the Ballast Water Management Convention, the Committee also approved a plan with specific arrangements for data gathering and analysis during the experience-building phase and approved guidance related to the form of the certificate, system and type approval process.

### *Hazardous and noxious substances*

In April 2018, the Legal Committee noted the latest States Parties to the 2010 Protocol to the International Convention on Liability and Compensation for Damage in Connection with the Carriage of Hazardous and Noxious Substances by Sea, 1996, namely Canada and Turkey (IMO, 2018g). To enter into force, the Convention requires accession by at least 12 States, representing at least 40 million tons of contributing cargo. As at 31 July 2018, it has been ratified by Canada, Norway and Turkey and the total of contributing cargo has reached 28.7 million tons or nearly 72 per cent of the amount required for its entry into force. Other States are encouraged to address, with a view to overcoming them, any practical issues and concerns related to implementing the Convention and to consider becoming Parties to it, to help cover a significant gap in the global liability and compensation framework. A comprehensive and robust international liability and compensation regime is in place with regard to oil pollution from tankers through the International Oil Pollution Compensation Fund regime, which includes the International Convention on Civil Liability for Oil Pollution Damage and its Protocol and the International Convention on the Establishment of an International Fund for Compensation for Oil Pollution Damage, 1971, and its 1992 and 2003 Protocols; and with regard to bunker oil pollution from ships other than tankers through the International Convention on Civil Liability for Bunker Oil Pollution Damage, 2001. However, at present, there is no international liability and compensation regime in place for hazardous and noxious substances that may cause significant personal injury and marine pollution (for an analytical overview of the international legal framework, see UNCTAD, 2012b, and UNCTAD, 2013).

#### *Marine litter*

In April 2018, the Marine Environment Protection Committee agreed to include a new item on its agenda to address the issue of marine plastic litter from shipping in the context of Sustainable Development Goal 14 (IMO, 2018e). Member States and international organizations were invited to submit proposals on the development of an action plan to the next session of the Committee. The issue of marine debris, plastics and microplastics in the oceans has been receiving increasing public attention and was the topic of focus at the seventeenth meeting of the United Nations Open-ended Informal Consultative Process on Oceans and the Law of the Sea in 2016 (United Nations, 2016). Marine debris in general, and plastics and microplastics in particular, are one of the greatest current environmental concerns, along with climate change, ocean acidification and the loss of biodiversity, which directly affect the sustainable development aspirations of developing States, in particular small island developing States, which, as custodians of vast areas of oceans and seas, face "an existential threat from and [are] disproportionately affected by the effects of pollution from plastics" (United Nations, 2016). Target 14.1 to, by 2025, prevent and significantly reduce marine pollution of all kinds, in particular from land-based activities, including marine debris and nutrient pollution, is particularly relevant in this context. Given the cross-cutting nature of the issue, other Goals are also relevant, including Goal 4 on education, Goal 6 on water and sanitation, Goal 12 on sustainable consumption and production patterns and Goal 15 on the sustainable use of terrestrial ecosystems.

## C. OTHER LEGAL AND REGULATORY DEVELOPMENTS AFFECTING TRANSPORTATION

### 1. Seafarers' issues

In April 2018, the Legal Committee highlighted the increased number of cases of abandonment of seafarers, as recorded in a joint IMO and International Labour Organization database; from 12–19 annual cases in 2011–2016, the number had risen to 55 cases in 2017 (IMO, 2018g). Shipowners in financial difficulty may abandon seafarers in ports far from home, leaving them without food, water, medical care, fuel or pay for months at a time. The 2014 amendments to the Maritime Labour Convention that entered into force in January 2017 make insurance to cover such abandonment, as well as claims for the death or long-term disability of seafarers, compulsory for shipowners. The worldwide population of seafarers serving on internationally trading merchant ships is estimated at 1,647,500, and most are from developing countries; China, the Philippines, Indonesia, the Russian Federation and Ukraine are estimated as the five leading seafarer supply countries (International Chamber of Shipping, 2017). The secretariats of IMO and the International Labour Organization were requested to consult on the inclusion in the database of information related to insurance for each new case and to prepare a list of competent authorities and organizations that could assist in resolving cases (IMO, 2018g). In addition, the Committee was advised of guidance being developed by the International Transport Workers' Federation and Seafarers' Rights International to support the implementation of the IMO and International Labour Organization guidelines on the fair treatment of seafarers in the event of a maritime accident, in view of the different approaches that States had taken in implementing the guidelines. The guidelines aim to ensure that seafarers are treated fairly following a maritime accident and during any investigation and detention by public authorities and that detention is for no longer than necessary. A comprehensive survey conducted by Seafarers' Rights International in 2011–2012 had suggested that the rights of seafarers as detailed in the guidelines were often subject to violation (IMO, 2018h).

### 2. Fraudulent registration

In the last few years, several member States have reported to the IMO secretariat cases of fraudulent use of their flags, with many illegally registered ships, some of which have been involved in illicit activities. In April 2018, the Legal Committee agreed that the fraudulent registration of ships needed to be addressed and that effective enforcement measures to discourage the practice and prevent ships with fraudulent registration from operating should be considered. The issue is complex, however, as it involves aspects of public international law and private law, and a multipronged approach is needed. The IMO secretariat was requested to conduct a study of cases received and provide information on the capability of the Global Integrated Shipping Information System of IMO to address the issue, potentially including contact points, sample certificates and a list of registries (IMO, 2018g). The consideration of measures to prevent unlawful practices associated with the fraudulent registration and registries of ships was included in the work programme of the Legal Committee, with a target completion date of 2021.

## 3. Legally binding instrument under the United Nations Convention on the Law of the Sea

Under this Convention, resources found in the seabed beyond the limits of national jurisdiction are to be used for the benefit of humanity as a whole, with particular consideration for the interests and needs of developing countries (article 140). However, the Convention does not include a provision on the use of marine genetic resources found in the water column, which are commercially valuable and hold considerable potential for the development of advanced pharmaceuticals. Their exploitation may, in the near future, become a promising activity in areas beyond the limits of national jurisdiction. In the absence of a specific international legal framework regulating related issues, negotiations have been ongoing since 2016 at the United Nations on key elements for an international legally binding instrument under this Convention on the conservation and sustainable use of marine biological diversity of areas beyond the limits of national jurisdiction. The outcome of the fourth meeting of the preparatory committee established in accordance with General Assembly resolution 69/292 of 19 June 2015, held in July 2017, included a number of elements recommended for consideration by the General Assembly in the elaboration of a text (UNCTAD, 2017a; see www.un.org/Depts/los/biodiversity/prepcom.htm). The General Assembly, in its resolution 72/249 adopted on 24 December 2017, decided to convene an intergovernmental conference under the auspices of the United Nations to consider the recommendations of the preparatory committee on the elements and to elaborate the text of an international legally binding instrument under the Convention. The first session is scheduled to be held from 4 to 17 September 2018.

**Table 5.1 Contracting States Parties to selected international conventions on maritime transport, as at 31 July 2018**

| Title of convention | Date of entry into force or conditions for entry into force | Contracting States |
|---|---|---|
| United Nations Convention on a Code of Conduct for Liner Conferences, 1974 | 6 October 1983 | Algeria, Bangladesh, Barbados, Belgium, Benin, Burkina Faso, Burundi, Cameroon, Cabo Verde, Central African Republic, Chile, China, Congo, Costa Rica, Côte d'Ivoire, Cuba, Czechia, Democratic Republic of the Congo, Egypt, Ethiopia, Finland, France, Gabon, Gambia, Ghana, Guatemala, Guinea, Guyana, Honduras, India, Indonesia, Iraq, Italy, Jamaica, Jordan, Kenya, Kuwait, Lebanon, Liberia, Madagascar, Malaysia, Mali, Mauritania, Mauritius, Mexico, Montenegro, Morocco, Mozambique, Niger, Nigeria, Norway, Pakistan, Peru, Philippines, Portugal, Qatar, Republic of Korea, Romania, Russian Federation, Saudi Arabia, Senegal, Serbia, Sierra Leone, Slovakia, Somalia, Spain, Sri Lanka, Sudan, Sweden, Togo, Trinidad and Tobago, Tunisia, United Republic of Tanzania, Uruguay, Venezuela (Bolivarian Republic of), Zambia (76) |
| United Nations Convention on the Carriage of Goods by Sea, 1978 (Hamburg Rules) | 1 November 1992 | Albania, Austria, Barbados, Botswana, Burkina Faso, Burundi, Cameroon, Chile, Czechia, Dominican Republic, Egypt, Gambia, Georgia, Guinea, Hungary, Jordan, Kazakhstan, Kenya, Lebanon, Lesotho, Liberia, Malawi, Morocco, Nigeria, Paraguay, Romania, Saint Vincent and the Grenadines, Senegal, Sierra Leone, Syrian Arab Republic, Tunisia, Uganda, United Republic of Tanzania, Zambia (34) |
| United Nations Convention on International Multimodal Transport of Goods, 1980 | Not yet in force – requires 30 Contracting Parties | Burundi, Chile, Georgia, Lebanon, Liberia, Malawi, Mexico, Morocco, Rwanda, Senegal, Zambia (11) |
| United Nations Convention on Conditions for Registration of Ships, 1986 | Not yet in force – requires 40 Contracting Parties with at least 25 per cent of the world's tonnage as per annex III to the Convention | Albania, Bulgaria, Côte d'Ivoire, Egypt, Georgia, Ghana, Haiti, Hungary, Iraq, Liberia, Libya, Mexico, Morocco, Oman, Syrian Arab Republic (15) |
| International Convention on Maritime Liens and Mortgages, 1993 | 5 September 2004 | Albania, Benin, Congo, Ecuador, Estonia, Lithuania, Monaco, Nigeria, Peru, Russian Federation, Spain, Saint Kitts and Nevis, Saint Vincent and the Grenadines, Serbia, Syrian Arab Republic, Tunisia, Ukraine, Vanuatu (18) |
| International Convention on Arrest of Ships, 1999 | 14 September 2011 | Albania, Algeria, Benin, Bulgaria, Congo, Ecuador, Estonia, Latvia, Liberia, Spain, Syrian Arab Republic (11) |

*Note:* For official status information, see the United Nations Treaty Collection, available at https://treaties.un.org, and UNCTAD, Conventions on commercial maritime law, available at http://unctad.org/en/Pages/DTL/TTL/Legal/Maritime-Conventions.aspx.

## D. STATUS OF CONVENTIONS

A number of international conventions in the field of maritime transport were prepared or adopted under the auspices of UNCTAD. Table 5.1 provides information on the status of ratification of each of these conventions as at 31 July 2018.

## E. OUTLOOK AND POLICY CONSIDERATIONS

Ongoing incidents against systems on board ships and in ports, which have significantly affected the maritime industry, highlight the importance of cybersecurity and cyberrisk management. At the international level, in addition to the IMO guidelines on maritime cyberrisk management adopted in 2017, an IMO resolution encourages administrations to ensure that cyberrisks are appropriately addressed in existing safety management systems, from 1 January 2021. This is the first compulsory deadline in the maritime industry related to cyberrisks and is an important step in protecting the maritime transportation system and the maritime industry from ever-increasing cybersecurity threats. In addition, the strategic plan for IMO adopted in 2017 recognizes the need to integrate new and emerging technologies into the regulatory framework for shipping, by balancing the benefits derived from such technologies "against safety and security concerns, the impact on the environment and on international trade facilitation, the potential costs to the industry and finally their impact on personnel, both on board and ashore" (IMO, 2017c). At the same time, the shipping industry is taking a proactive approach to incorporating cyberrisk management in its safety culture, to prevent the occurrence of any serious incidents. Relevant guidance has been and continues to be developed by classification societies and other industry associations, as well as by individual States, providing practical recommendations on maritime cyberrisk management and including information on insurance issues.

With regard to distributed ledger technology such as blockchain, at present, many initiatives and partnerships are emerging and proliferating, including in the shipping industry. Greater numbers of stakeholders are exploring its utilization, including for digitalizing and automating paper filing, documents, smart contracts and insurance policies, to save time and reduce costs in the clearance and movement of cargo. Such initiatives need to be interoperable, as competition between them in a bid to make a specific technology the chosen standard for the industry may be detrimental for shipping. In addition, blockchain promises secure transactions yet, according to some specialists, may not be as secure as generally anticipated. The use of blockchain may help solve some security issues but may also lead to new, potentially more complex security challenges. UNCTAD has also noted related general concerns about the mix of benefits and risks of digitalization as a disruptive technology. Many developing countries, in particular the least developed countries, may be inadequately prepared to capture the opportunities and benefits emerging from digitalization, and there may be a risk that this could lead to increased polarization and widening income inequalities.

The development and use of autonomous ships present numerous benefits, yet it is unclear whether this advance in technology will be fully accepted by Governments and by the traditionally conservative maritime industry. There are concerns about the safety and security of operations and the reliability of autonomous ships, as well as the diminishing role of and loss of jobs for seafarers, the majority of which are from developing countries. In addition, the use of autonomous ships poses a number of legal and regulatory compliance-related issues that need to be considered and addressed. Conducting regulatory reviews and scoping exercises are therefore of particular importance. Similar issues arise in connection with the use of drones, which has the potential to generate important benefits and may be encouraged; at the same time, the applicable regulatory framework needs to be further studied and developed.

Complementing international efforts to address greenhouse gas emissions – including under the Paris Agreement and the 2030 Agenda, in particular Goal 13 – in 2018, an important achievement at IMO related to the determination of the fair share of emissions reduction by international shipping was the adoption of an initial strategy on the reduction of greenhouse gas emissions from ships, according to which total annual greenhouse gas emissions should be reduced by at least 50 per cent by 2050, compared with 2008. The strategy identifies candidate short-term, midterm and long-term further measures, with possible timelines and their impacts on States, stating that specific attention should be paid to the needs of developing countries, in particular the least developed countries and small island developing States. It also identifies supportive measures, including capacity-building, technical cooperation and research and development.

The implementation of technical and operational measures, as well as the development of innovative technologies for ships, are ongoing. Amendments to the International Convention for the Prevention of Pollution from Ships have entered into force that make data collection systems for the fuel oil consumption of ships of 5,000 gross tons and above mandatory, with data collection from 1 January 2019. The data must be reported to the flag State after the end of each calendar year and subsequently transferred to the IMO database. With regard to ship-source air pollution, the global limit of 0.5 per cent on sulphur in fuel oil outside emission control areas will come into effect on 1 January 2020. The consistent implementation of the limit for all ships is expected to bring positive results for human health and the environment. Guidelines to support the implementation of the limit are being prepared by IMO. It is important for shipowners and operators to continue to consider and adopt various relevant strategies, including installing scrubbers and switching to liquefied natural gas and other low-sulphur fuels.

Given the importance of implementing and effectively enforcing strong international environmental regulations and in the light of the policy objectives under Sustainable Development Goal 14, developed and developing countries are encouraged to consider becoming parties to relevant international conventions for the prevention and control of marine pollution as a matter of priority. The widespread adoption and implementation of international conventions addressing liability and compensation for shipsource pollution, such as the International Convention on Liability and Compensation for Damage in Connection with the Carriage of Hazardous and Noxious Substances by Sea, is also desirable in view of the significant gaps that remain in the international legal framework.

# REFERENCES

All About Shipping (2018). Cyberrisk exercises marine insurers. 7 February.

Allianz Global Corporate and Specialty (2017). *Safety and Shipping Review 2017*. Munich.

Baird Maritime (2018). Norway investigates offshore drones delivering cargo. 22 February.

BIMCO, Cruise Lines International Association, International Chamber of Shipping, International Association of Dry Cargo Shipowners, International Association of Independent Tanker Owners, Oil Companies International Marine Forum and International Union of Marine Insurance (2017). The guidelines on cybersecurity on board ships, version 2.0. Available at www.bimco.org/products/publications/free/cyber-security.

Bloomberg (2017). This robot ship experiment could disrupt the global shipping industry. 23 August.

*Combined Transport Magazine* (2016). Secure data exchange across supply chains: Blockchain and electronic data interchange. 9 November.

Comité Maritime International (2017). International working group position paper on unmanned ships and the international regulatory framework. Available at http://comitemaritime.org/work/unmanned-ships/.

Danish Maritime Authority (2017). *Analysis of Regulatory Barriers to the Use of Autonomous Ships*. Final report. Available at www.dma.dk/Vaekst/autonomeskibe/Pages/Foranalyse-af-autonome-skibe.aspx.

DNV GL (2017). DNV GL carries out its first offshore drone survey. 3 August.

DNV GL (2018). The ReVolt: A new inspirational ship concept.

Economic Commission for Europe (1996). Recommendation 25: Use of the United Nations Electronic Data Interchange for administration, commerce and transport. TRADE/WP.4/R.1079/Rev.1. Geneva.

Fairplay (2017). Insurance industry expresses concerns over autonomous vessels. 20 November.

Fast Company (2017). A start-up's plan to cut air freight costs in half with 777-size drones. 27 March.

Guardtime (2017). [Ernst and Young], Guardtime and industry participants launch the world's first marine insurance blockchain platform. 4 September. Available at https://guardtime.com/blog/ey-guardtime-world-s-first-marine-insurance-blockchain-platform.

IMO (2009). *Second IMO Greenhouse Gas Study 2009*. London.

IMO (2013). Report of the Marine Environment Protection Committee on its sixty-fifth session. MEPC 65/22. London. 24 May.

IMO (2014). *Third IMO Greenhouse Gas Study 2014*. London.

IMO (2016). Report of the Marine Environment Protection Committee on its seventieth session. MEPC 70/18. London. 11 November.

IMO (2017a). Report of the Maritime Safety Committee on its ninety-eighth session. MSC 98/23. London. 28 June.

IMO (2017b). Guidelines on maritime cyberrisk management. MSC-FAL.1/Circ.3. London. 5 July.

IMO (2017c). Strategic plan for the Organization for the six-year period 2018 to 2023. A.1110(30). London. 8 December.

IMO (2017d). Report of the first meeting of the Intersessional Working Group on Reduction of Greenhouse Gas Emissions from Ships. MEPC 71/WP.5. London.

IMO (2017e). Report of the second meeting of the Intersessional Working Group on Reduction of Greenhouse Gas Emissions from Ships. MEPC 72/7. London. 3 November.

IMO (2017f). IMO moves ahead with oceans and climate change agenda. 11 July. Available at www.imo.org/en/MediaCentre/PressBriefings/Pages/17-MEPC-71.aspx.

IMO (2018a). [International Safety Management] Code and guidelines on implementation of the [International Safety Management] Code. Available at www.imo.org/en/OurWork/HumanElement/SafetyManagement/Pages/ISMCode.aspx.

IMO (2018b). Proposal for a regulatory scoping exercise and gap analysis with respect to maritime autonomous surface ships. LEG 105/11/1. London. 30 January.

IMO (2018c). Regulatory scoping exercise for the use of maritime autonomous surface ships. Comments on the regulatory scoping exercise. MSC 99/5. London.

IMO (2018d). Report of the Maritime Safety Committee on its ninety-ninth session. MSC 99/22. London. 5 June.

IMO (2018e). Report of the Marine Environment Protection Committee on its seventy-second session. MEPC 72/17. London. 3 May.

IMO (2018f). Report of the Working Group on Reduction of Greenhouse Gas Emissions from Ships. MEPC 72/WP.7. London. 12 April.

IMO (2018g). Report of the Legal Committee on the work of its 105th session. LEG 105/14. London. 1 May.

IMO (2018h). Legal Committee, 105th session, 23–25 April 2018. 25 April. Available at www.imo.org/en/MediaCentre/MeetingSummaries/Legal/Pages/LEG-105th-session.aspx.

*Independent* (2018). Cleaner shipping fuels could prevent hundreds of thousands of emissions-related deaths, finds new study. 6 February.

International Chamber of Shipping (2017). Global supply and demand for seafarers. Available at www.ics-shipping.org/shipping-facts/shipping-and-world-trade/global-supply-and-demand-for-seafarers.

JOC.com (2018). Blockchain success in shipping hinges on standardization. 27 March.

Kongsberg (2017). Bourbon joins Automated Ships Ltd. and Kongsberg to deliver ground-breaking autonomous offshore support vessel prototype. 11 July.

Lloyd's List (2017). [Hyundai Merchant Marine] completes pilot blockchain voyage with reefer-laden box ship. 7 September.

Maersk (2018). Maersk and IBM to form joint venture applying blockchain to improve global trade and digitize supply chains. 16 January.

Marine Electronics and Communications (2018a). Blockchain is not the silver bullet for cybersecurity. 9 March.

Marine Electronics and Communications (2018b). More to autonomous technology than just unmanned ships. 28 March.

Marine Log (2017). Zim completes pilot of blockchain-based paperless bills of lading. 21 November.

Marine Log (2018). Naval Dome cybersecurity system completes box ship pilot testing. 5 February.

Rolls-Royce (2017). Rolls-Royce joins forces with Google Cloud to help make autonomous ships a reality. 3 October.

Rolls-Royce (2018). Rolls-Royce offers ship navigators a bird's-eye view with Intelligent Awareness game changer. 6 March.

Splash 247 (2018). Maersk successfully pilots first marine insurance blockchain platform. 25 May.

Stopford M (2009). *Maritime Economics*. 4th ed. Routledge. Abingdon, United Kingdom.

SUAS News (2017). Martek Marine named on world's biggest ever €67 million maritime drone contract. 17 March.

The Conversation (2018a). How blockchain is strengthening tuna traceability to combat illegal fishing. 21 January.

The Conversation (2018b). Who's to blame when driverless cars have an accident? 20 March.

*The Guardian* (2017). WannaCry, Petya, NotPetya: How ransomware hit the big time in 2017. 30 December.

*The Maritime Executive* (2017). Wilhelmsen launches delivery drone service at Nor Shipping. 19 May.

*The Maritime Executive* (2018). Dutch shipowner orders electric inland barges. 22 January.

UASweekly.com (2018). SSE chooses Martek Aviation to inspect 683 wind turbines. 26 January.

UNCTAD (2003). The use of transport documents in international trade. Available at http://unctad.org/en/Pages/DTL/TTL/Legal/Carriage-of-Goods.aspx.

UNCTAD (2011a). *Review of Maritime Transport 2011* (United Nations publication. Sales No. E.11.II.D.4. New York and Geneva).

UNCTAD (2011b). The 2004 Ballast Water Management Convention – with international acceptance growing, the Convention may soon enter into force. In: Transport Newsletter No. 50.

UNCTAD (2012a). *Review of Maritime Transport 2012* (United Nations publication. Sales No. E.12.II.D.17. New York and Geneva).

UNCTAD (2012b). *Liability and Compensation for Ship-source Oil Pollution: An Overview of the International Legal Framework for Oil Pollution Damage from Tankers* (United Nations publication. New York and Geneva).

UNCTAD (2013). *Review of Maritime Transport 2013*. (United Nations publication. Sales No. E.13.II.D.9. New York and Geneva).

UNCTAD (2015). The International Ballast Water Management Convention 2004 is set to enter into force in 2016. In: Transport and Trade Facilitation Newsletter No. 68.

UNCTAD (2017a). *Review of Maritime Transport 2017* (United Nations publication. Sales No. E.17.II.D.10. New York and Geneva).

UNCTAD (2017b). *Information Economy Report 2017: Digitalization, Trade and Development* (United Nations publication. Sales No. E.17.II.D.8. New York and Geneva).

UNCTAD (2017c). *Trade and Development Report 2017: Beyond Austerity – Towards a Global New Deal* (United Nations publication. Sales No. E.17.II.D.5. New York and Geneva).

United Nations (2016). Report on the work of the United Nations Open-ended Informal Consultative Process on Oceans and the Law of the Sea at its seventeenth meeting. A/71/204. New York. 25 July.

Venture Beat (2017). Blockchain's brilliant approach to cybersecurity. 22 January.

Wärtsilä (2018). World's first autodocking installation successfully tested by Wärtsilä. 26 April.

ZD Net (2018). NonPetya ransomware forced Maersk to reinstall 4,000 servers, 45,000 [personal computers]. 26 January.